Science 900
Teacher's Guide

CONTENTS

PART I
Curriculum Overview

PART II
LIFEPAC Management

PART III
Teacher Notes

PART IV
Alternate Tests with Keys

PART V
Answer Keys - Self Test Keys - LIFEPAC Test Keys

Revision Editor: Alan Christopherson, M.S.

Alpha Omega Publications
a division of:
Bridgestone Multimedia Group

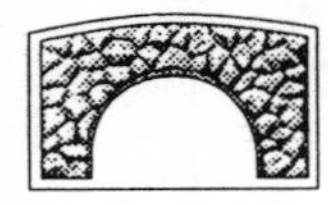

300 North McKemy Avenue, Chandler, Arizona 85226-2618

CURRICULUM

OVERVIEW

SCIENCE

Curriculum Overview
Grades 1–12

Science LIFEPAC Overview

	Grade 1	Grade 2	Grade 3
LIFEPAC 1	YOU LEARN WITH YOUR EYES • Name and group some colors • Name and group some shapes • Name and group some sizes • Help from what you see	THE LIVING AND NONLIVING • What God created • Rock and seed experiment • God-made objects • Man-made objects	YOU GROW AND CHANGE • Air we breathe • Food for the body • Exercise and rest • You are different
LIFEPAC 2	YOU LEARN WITH YOUR EARS • Sounds of nature and people • How sound moves • Sound with your voice • You make music	PLANTS • How are plants alike • Habitats of plants • Growth of plants • What plants need	PLANTS • Plant parts • Plant growth • Seeds and bulbs • Stems and roots
LIFEPAC 3	MORE ABOUT YOUR SENSES • Sense of smell • Sense of taste • Sense of touch • Learning with my senses	ANIMALS • How are animals alike • How are animals different • What animals need • Noah and the ark	ANIMAL GROWTH AND CHANGE • The environment changes • Animals are different • How animals grow • How animals change
LIFEPAC 4	ANIMALS • What animals eat • Animals for food • Animals for work • Pets to care for	YOU • How are people alike • How are you different • Your family • Your health	YOU ARE WHAT YOU EAT • Food helps your body • Junk foods • Food groups • Good health habits
LIFEPAC 5	PLANTS • Big and small plants • Special plants • Plants for food • House plants	PET AND PLANT CARE • Learning about pets • Caring for pets • Learning about plants • Caring for plants	PROPERTIES OF MATTER • Robert Boyle • States of matter • Physical changes • Chemical changes
LIFEPAC 6	GROWING UP HEALTHY • How plants and animals grow • How your body grows • Eating and sleeping • Exercising	YOUR FIVE SENSES • Your eye • You can smell and hear • Your taste • You can feel	SOUNDS AND YOU • Making sounds • Different sounds • How sounds move • How sounds are heard
LIFEPAC 7	GOD'S BEAUTIFUL WORLD • Types of land • Water places • The weather • Seasons	PHYSICAL PROPERTIES • Colors • Shapes • Sizes • How things feel	TIMES AND SEASONS • The earth rotates • The earth revolves • Time changes • Seasons change
LIFEPAC 8	ALL ABOUT ENERGY • God gives energy • We use energy • Ways to make energy • Ways to save energy	OUR NEIGHBORHOOD • Things not living • Things living • Harm to our world • Caring for our world	ROCKS AND THEIR CHANGES • Forming rocks • Changing rocks • Rocks for buildings • Rock collecting
LIFEPAC 9	MACHINES AROUND YOU • Simple levers • Simple wheels • Inclined planes • Using machines	CHANGES IN OUR WORLD • Seasons • Change in plants • God's love never changes • God's Word never changes	HEAT ENERGY • Sources of heat • Heat energy • Moving heat • Benefits and problems of heat
LIFEPAC 10	WONDERFUL WORLD OF SCIENCE • Using your senses • Using your mind • You love yourself • You love the world	LOOKING AT OUR WORLD • Living things • Nonliving things • Caring for our world • Caring for ourselves	PHYSICAL CHANGES • Change in man • Change in plants • Matter and time • Sound and energy

	Grade 4	Grade 5	Grade 6
LIFEPAC 1	PLANTS • Plants and living things • Using plants • Parts of plants • The function of plants	CELLS • Cell composition • Plant and animal cells • Life of cells • Growth of cells	PLANT SYSTEMS • Parts of a plant • Systems of photosynthesis • Transport systems • Regulatory systems
LIFEPAC 2	ANIMALS • Animal structures • Animal behavior • Animal instincts • Man protects animals	PLANTS: LIFE CYCLES • Seed producing plants • Spore producing plants • One-celled plants • Classifying plants	ANIMAL SYSTEMS • Digestive system • Excretory system • Skeletal system • Diseases
LIFEPAC 3	MAN'S ENVIRONMENT • Resources • Balance in nature • Communities • Conservation and preservation	ANIMALS: LIFE CYCLES • Invertebrates • Vertebrates • Classifying animals • Relating function and structure	PLANT AND ANIMAL BEHAVIOR • Animal behavior • Plant behavior • Plant-animal interaction • Balance in nature
LIFEPAC 4	MACHINES • Work and energy • Simple machines • Simple machines together • Complex machines	BALANCE IN NATURE • Needs of life • Dependence on others • Prairie life • Stewardship of nature	MOLECULAR GENETICS • Reproduction • Inheritance • DNA and mutations • Mendel's work
LIFEPAC 5	ELECTRICITY AND MAGNETISM • Electric current • Electric circuits • Magnetic materials • Electricity and magnets	TRANSFORMATION OF ENERGY • Work and energy • Heat energy • Chemical energy • Energy sources	CHEMICAL STRUCTURE • Nature of matter • Periodic Table • Diagrams of atoms • Acids and bases
LIFEPAC 6	CHANGES IN MATTER • Properties of water • Properties of matter • Molecules and atoms • Elements	RECORDS IN ROCK: THE FLOOD • The Biblical account • Before the flood • The flood • After the flood	LIGHT AND SOUND • Sound waves • Light waves • The visible spectrum • Colors
LIFEPAC 7	WEATHER • Causes of weather • Forces of weather • Observing weather • Weather instruments	RECORDS IN ROCK: FOSSILS • Fossil types • Fossil location • Identifying fossils • Reading fossils	MOTION AND ITS MEASUREMENT • Definition of force • Rate of doing work • Laws of motion • Change in motion
LIFEPAC 8	THE SOLAR SYSTEM • Our solar system • The big universe • Sun and planets • Stars and space	RECORDS IN ROCK: GEOLOGY • Features of the earth • Rock of the earth • Forces of the earth • Changes in the earth	SPACESHIP EARTH • Shape of the earth • Rotation and revolution • Eclipses • The solar system
LIFEPAC 9	THE PLANET EARTH • The atmosphere • The hydrosphere • The lithosphere • Rotation and revolution	CYCLES IN NATURE • Properties of matter • Changes in matter • Natural cycles • God's order	SUN AND OTHER STARS • The sun • Investigating stars • Common stars • Constellations
LIFEPAC 10	GOD'S CREATION • Earth and solar system • Matter and weather • Using nature • Conservation	LOOK AHEAD • Plant and animal life • Balance in nature • Biblical records • Records of rock	THE EARTH AND THE UNIVERSE • Plant systems • Animal systems • Physics and chemistry • The earth and stars

Science LIFEPAC Overview

	Grade 7	Grade 8	Grade 9
LIFEPAC 1	WHAT IS SCIENCE • Tools of a scientist • Methods of a scientist • Work of a scientist • Careers in science	SCIENCE AND SOCIETY • Definition of science • History of science • Science today • Science tomorrow	OUR ATOMIC WORLD • Structure of matter • Radioactivity • Atomic nuclei • Nuclear energy
LIFEPAC 2	PERCEIVING THINGS • History of the metric system • Metric units • Advantages of the metric system • Graphing data	STRUCTURE OF MATTER I • Properties of matter • Chemical properties of matter • Atoms and molecules • Elements, compounds, & mixtures	VOLUME, MASS, AND DENSITY • Measure of matter • Volume • Mass • Density
LIFEPAC 3	EARTH IN SPACE I • Ancient stargazing • Geocentric Theory • Copernicus • Tools of astronomy	STRUCTURE OF MATTER II • Changes in matter • Acids • Bases • Salts	PHYSICAL GEOLOGY • Earth structures • Weathering and erosion • Sedimentation • Earth movements
LIFEPAC 4	EARTH IN SPACE II • Solar energy • Planets of the sun • The moon • Eclipses	HEALTH AND NUTRITION • Foods and digestion • Diet • Nutritional diseases • Hygiene	HISTORICAL GEOLOGY • Sedimentary rock • Fossils • Crustal changes • Measuring time
LIFEPAC 5	THE ATMOSPHERE • Layers of the atmosphere • Solar effects • Natural cycles • Protecting the atmosphere	ENERGY I • Kinetic and potential energy • Other forms of energy • Energy conversions • Entropy	BODY HEALTH I • Microorganisms • Bacterial infections • Viral infections • Other infections
LIFEPAC 6	WEATHER • Elements of weather • Air masses and clouds • Fronts and storms • Weather forecasting	ENERGY II • Magnetism • Current and static electricity • Using electricity • Energy sources	BODY HEALTH II • Body defense mechanisms • Treating disease • Preventing disease • Community health
LIFEPAC 7	CLIMATE • Climate and weather • Worldwide climate • Regional climate • Local climate	MACHINES I • Measuring distance • Force • Laws of Newton • Work	ASTRONOMY • Extent of the universe • Constellations • Telescopes • Space explorations
LIFEPAC 8	HUMAN ANATOMY I • Cell structure and function • Skeletal and muscle systems • Skin • Nervous system	MACHINES II • Friction • Levers • Wheels and axles • Inclined planes	OCEANOGRAPHY • History of oceanography • Research techniques • Geology of the ocean • Properties of the ocean
LIFEPAC 9	HUMAN ANATOMY II • Respiratory system • Circulatory system • Digestive system • Endocrine system	BALANCE IN NATURE • Photosynthesis • Food • Natural cycles • Balance in nature	SCIENCE AND TOMORROW • The land • Waste and ecology • Industry and energy • New frontiers
LIFEPAC 10	CAREERS IN SCIENCE • Scientists at work • Astronomy • Meteorology • Medicine	SCIENCE AND TECHNOLOGY • Basic science • Physical science • Life science • Vocations in science	SCIENTIFIC APPLICATIONS • Measurement • Practical health • Geology and astronomy • Solving problems

Grade 10	Grade 11	Grade 12	
TAXONOMY • History of taxonomy • Binomial nomenclature • Classification • Taxonomy	INTRODUCTION TO CHEMISTRY • Metric units and instrumentation • Observation and hypothesizing • Scientific notation • Careers in chemistry	KINEMATICS • Scalars and vectors • Length measurement • Acceleration • Fields and models	LIFEPAC 1
BASIS OF LIFE • Elements and molecules • Properties of compounds • Chemical reactions • Organic compounds	BASIC CHEMICAL UNITS • Alchemy • Elements • Compounds • Mixtures	DYNAMICS • Newton's Laws of Motion • Gravity • Circular motion • Kepler's Laws of Motion	LIFEPAC 2
MICROBIOLOGY • The microscope • Protozoan • Algae • Microorganisms	GASES AND MOLES • Kinetic theory • Gas laws • Combined gas law • Moles	WORK AND ENERGY • Mechanical energy • Conservation of energy • Power and efficiency • Heat energy	LIFEPAC 3
CELLS • Cell theories • Examination of the cell • Cell design • Cells in organisms	ATOMIC MODELS • Historical models • Modern atomic structure • Periodic Law • Nuclear reactions	WAVES • Energy transfers • Reflection and refraction of waves • Diffraction and interference • Sound waves	LIFEPAC 4
PLANTS: GREEN FACTORIES • The plant cell • Anatomy of the plant • Growth and function of plants • Plants and people	CHEMICAL FORMULAS • Ionic charges • Electronegativity • Chemical bonds • Molecular shape	LIGHT • Speed of light • Mirrors • Lenses • Models of light	LIFEPAC 5
HUMAN ANATOMY AND PHYSIOLOGY • Digestive and excretory system • Respiratory and circulatory system • Skeletal and muscular system • Body control systems	CHEMICAL REACTIONS • Detecting reactions • Energy changes • Reaction rates • Equilibriums	STATIC ELECTRICITY • Nature of charges • Transfer of charges • Electric fields • Electric potential	LIFEPAC 6
INHERITANCE • Gregor Mendel's experiments • Chromosomes and heredity • Molecular genetics • Human genetics	EQUILIBRIUM SYSTEMS • Solutions • Solubility equilibriums • Acid-base equilibriums • Redox equilibriums	CURRENT ELECTRICITY • Electromotive force • Electron flow • Resistance • Circuits	LIFEPAC 7
CELL DIVISION & REPRODUCTION • Mitosis and meiosis • Asexual reproduction • Sexual reproduction • Plant reproduction	HYDROCARBONS • Organic compounds • Carbon atoms • Carbon bonds • Saturated and unsaturated	MAGNETISM • Fields • Forces • Electromagnetism • Electron beams	LIFEPAC 8
ECOLOGY & ENERGY • Ecosystems • Communities and habitats • Pollution • Energy	CARBON CHEMISTRY • Saturated and unsaturated • Reaction types • Oxygen groups • Nitrogen groups	ATOMIC AND NUCLEAR PHYSICS • Electromagnetic radiation • Quantum theory • Nuclear theory • Nuclear reaction	LIFEPAC 9
APPLICATIONS OF BIOLOGY • Principles of experimentation • Principles of reproduction • Principles of life • Principles of ecology	ATOMS TO HYDROCARBONS • Atoms and molecules • Chemical bonding • Chemical systems • Organic chemistry	KINEMATICS TO NUCLEAR PHYSICS • Mechanics • Wave motion • Electricity • Modern physics	LIFEPAC 10

LIFEPAC

MANAGEMENT

STRUCTURE OF THE LIFEPAC CURRICULUM

The LIFEPAC curriculum is conveniently structured to provide one teacher handbook containing teacher support material with answer keys and ten student worktexts for each subject at grade levels two through twelve. The worktext format of the LIFEPACs allows the student to read the textual information and complete workbook activities all in the same booklet. The easy to follow LIFEPAC numbering system lists the grade as the first number(s) and the last two digits as the number of the series. For example, the Language Arts LIFEPAC at the 6th grade level, 5th book in the series would be LA 605.

Each LIFEPAC is divided into 3 to 5 sections and begins with an introduction or overview of the booklet as well as a series of specific learning objectives to give a purpose to the study of the LIFEPAC. The introduction and objectives are followed by a vocabulary section which may be found at the beginning of each section at the lower levels, at the beginning of the LIFEPAC in the middle grades, or in the glossary at the high school level. Vocabulary words are used to develop word recognition and should not be confused with the spelling words introduced later in the LIFEPAC. The student should learn all vocabulary words before working the LIFEPAC sections to improve comprehension, retention, and reading skills.

Each activity or written assignment has a number for easy identification, such as 1.1. The first number corresponds to the LIFEPAC section and the number to the right of the decimal is the number of the activity.

Teacher checkpoints, which are essential to maintain quality learning, are found at various locations throughout the LIFEPAC. The teacher should check 1) neatness of work and penmanship, 2) quality of understanding (tested with a short oral quiz), 3) thoroughness of answers (complete sentences and paragraphs, correct spelling, etc.), 4) completion of activities (no blank spaces), and 5) accuracy of answers as compared to the answer key (all answers correct).

The self test questions are also number coded for easy reference. For example, 2.015 means that this is the 15th question in the self test of Section II. The first number corresponds to the LIFEPAC section, the zero indicates that it is a self test question, and the number to the right of the zero the question number.

The LIFEPAC test is packaged at the centerfold of each LIFEPAC. It should be removed and put aside before giving the booklet to the student for study.

Answer and test keys have the same numbering system as the LIFEPACs and appear at the back of this handbook. The student may be given access to the answer keys (not the test keys) under teacher supervision so that he can score his own work.

A thorough study of the Curriculum Overview by the teacher before instruction begins is essential to the success of the student. The teacher should become familiar with expected skill mastery and understand how these grade level skills fit into the overall skill development of the curriculum. The teacher should also preview the objectives that appear at the beginning of each LIFEPAC for additional preparation and planning.

TEST SCORING and GRADING

Answer keys and test keys give examples of correct answers. They convey the idea, but the student may use many ways to express a correct answer. The teacher should check for the essence of the answer, not for the exact wording. Many questions are high level and require thinking and creativity on the part of the student. Each answer should be scored based on whether or not the main idea written by the student matches the model example. "Any Order" or "Either Order" in a key indicates that no particular order is necessary to be correct.

Most self tests and LIFEPAC tests at the lower elementary levels are scored at 1 point per answer; however, the upper levels may have a point system awarding 2 to 5 points for various answers or questions. Further, the total test points will vary; they may not always equal 100 points. They may be 78, 85, 100, 105, etc.

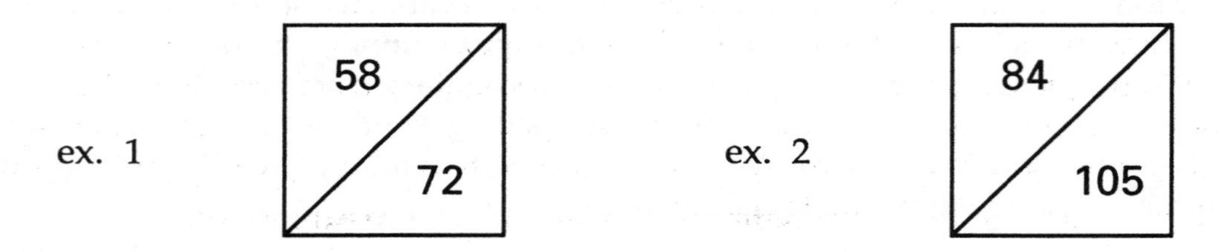

A score box similar to ex.1 above is located at the end of each self test and on the front of the LIFEPAC test. The bottom score, 72, represents the total number of points possible on the test. The upper score, 58, represents the number of points your student will need to receive an 80% or passing grade. If you wish to establish the exact percentage that your student has achieved, find the total points of his correct answers and divide it by the bottom number (in this case 72.) For example, if your student has a point total of 65, divide 65 by 72 for a grade of 90%. Referring to ex. 2, on a test with a total of 105 possible points, the student would have to receive a minimum of 84 correct points for an 80% or passing grade. If your student has received 93 points, simply divide the 93 by 105 for a percentage grade of 86%. Students who receive a score below 80% should review the LIFEPAC and retest using the appropriate Alternate Test found in the Teacher's Guide.

The following is a guideline to assign letter grades for completed LIFEPACs based on a maximum total score of 100 points.

LIFEPAC Test = 60% of the Total Score (or percent grade)
Self Test = 25% of the Total Score (average percent of self tests)
Reports = 10% or 10* points per LIFEPAC
Oral Work = 5% or 5* points per LIFEPAC
*Determined by the teacher's subjective evaluation of the student's daily work.

Example:

LIFEPAC Test Score	=	92%	92 x .60	=	55 points
Self Test Average	=	90%	90 x .25	=	23 points
Reports				=	8 points
Oral Work				=	4 points
TOTAL POINTS				=	90 points

Grade Scale based on point system:

100	–	94	=	A
93	–	86	=	B
85	–	77	=	C
76	–	70	=	D
Below		70	=	F

TEACHER HINTS and STUDYING TECHNIQUES

LIFEPAC Activities are written to check the level of understanding of the preceding text. The student may look back to the text as necessary to complete these activities; however, a student should never attempt to do the activities without reading (studying) the text first. Self tests and LIFEPAC tests are never open book tests.

Language arts activities (skill integration) often appear within other subject curriculum. The purpose is to give the student an opportunity to test his skill mastery outside of the context in which it was presented.

Writing complete answers (paragraphs) to some questions is an integral part of the LIFEPAC Curriculum in all subjects. This builds communication and organization skills, increases understanding and retention of ideas, and helps enforce good penmanship. Complete sentences should be encouraged for this type of activity. Obviously, single words or phrases do not meet the intent of the activity, since multiple lines are given for the response.

Review is essential to student success. Time invested in review where review is suggested will be time saved in correcting errors later. Self tests, unlike the section activities, are closed book. This procedure helps to identify weaknesses before they become too great to overcome. Certain objectives from self tests are cumulative and test previous sections; therefore, good preparation for a self test must include all material studied up to that testing point.

The following procedure checklist has been found to be successful in developing good study habits in the LIFEPAC curriculum.

1. Read the introduction and Table of Contents.
2. Read the objectives.
3. Recite and study the entire vocabulary (glossary) list.
4. Study each section as follows:
    a. Read the introduction and study the section objectives.
    b. Read all the text for the entire section, but answer none of the activities.
    c. Return to the beginning of the section and memorize each vocabulary word and definition.
    d. Reread the section, complete the activities, check the answers with the answer key, correct all errors, and have the teacher check.
    e. Read the self test but do not answer the questions.
    f. Go to the beginning of the first section and reread the text and answers to the activities up to the self test you have not yet done.
    g. Answer the questions to the self test without looking back.
    h. Have the self test checked by the teacher.
    i. Correct the self test and have the teacher check the corrections.
    j. Repeat steps a–i for each section.

5. Use the SQ3R* method to prepare for the LIFEPAC test.
6. Take the LIFEPAC test as a closed book test.
7. LIFEPAC tests are administered and scored under direct teacher supervision. Students who receive scores below 80% should review the LIFEPAC using the SQ3R* study method and take the Alternate Test located in the Teacher Handbook. The final test grade may be the grade on the Alternate Test or an average of the grades from the original LIFEPAC test and the Alternate Test.

   *SQ3R: Scan the whole LIFEPAC,
   Question yourself on the objectives,
   Read the whole LIFEPAC again,
   Recite through an oral examination, and
   Review weak areas.

GOAL SETTING and SCHEDULES

Each school must develop its own schedule, because no single set of procedures will fit every situation. The following is an example of a daily schedule that includes the five LIFEPAC subjects as well as time slotted for special activities.

Possible Daily Schedule

8:15	–	8:25	Pledges, prayer, songs, devotions, etc.
8:25	–	9:10	Bible
9:10	–	9:55	Language Arts
9:55	–	10:15	Recess (juice break)
10:15	–	11:00	Mathematics
11:00	–	11:45	Social Studies
11:45	–	12:30	Lunch, recess, quiet time
12:30	–	1:15	Science
1:15	–		Drill, remedial work, enrichment*

*Enrichment: Computer time, physical education, field trips, fun reading, games and puzzles, family business, hobbies, resource persons, guests, crafts, creative work, electives, music appreciation, projects.

Basically, two factors need to be considered when assigning work to a student in the LIFEPAC curriculum.

The first is time. An average of 45 minutes should be devoted to each subject, each day. Remember, this is only an average. Because of extenuating circumstances a student may spend only 15 minutes on a subject one day and the next day spend 90 minutes on the same subject.

The second factor is the number of pages to be worked in each subject. A single LIFEPAC is designed to take 3 to 4 weeks to complete. Allowing about 3-4 days for LIFEPAC introduction, review, and tests, the student has approximately 15 days to complete the LIFEPAC pages. Simply take the number of pages in the LIFEPAC, divide it by 15 and you will have the number of pages that must be completed on a daily basis to keep the student on schedule. For example, a LIFEPAC containing 45 pages will require 3 completed pages per day. Again, this is only an average. While working a 45 page LIFEPAC, the student may complete only 1 page the first day if the text has a lot of activities or reports, but go on to complete 5 pages the next day.

Long range planning requires some organization. Because the traditional school year originates in the early fall of one year and continues to late spring of the following year, a calendar should be devised that covers this period of time. Approximate beginning and completion dates can be noted on the calendar as well as special occasions such as holidays, vacations and birthdays. Since each LIFEPAC takes 3-4 weeks or eighteen days to complete, it should take about 180 school days to finish a set of ten LIFEPACs. Starting at the beginning school date, mark off eighteen school days on the calendar and that will become the targeted completion date for the first LIFEPAC. Continue marking the calendar until you have established dates for the remaining nine LIFEPACs making adjustments for previously noted holidays and vacations. If all five subjects are being used, the ten established target dates should be the same for the LIFEPACs in each subject.

FORMS

The sample weekly lesson plan and student grading sheet forms are included in this section as teacher support materials and may be duplicated at the convenience of the teacher.

The student grading sheet is provided for those who desire to follow the suggested guidelines for assignment of letter grades found on page 3 of this section. The student's self test scores should be posted as percentage grades. When the LIFEPAC is completed the teacher should average the self test grades, multiply the average by .25 and post the points in the box marked self test points. The LIFEPAC percentage grade should be multiplied by .60 and posted. Next, the teacher should award and post points for written reports and oral work. A report may be any type of written work assigned to the student whether it is a LIFEPAC or additional learning activity. Oral work includes the student's ability to respond orally to questions which may or may not be related to LIFEPAC activities or any type of oral report assigned by the teacher. The points may then be totaled and a final grade entered along with the date that the LIFEPAC was completed.

The Student Record Book which was specifically designed for use with the Alpha Omega curriculum provides space to record weekly progress for one student over a nine week period as well as a place to post self test and LIFEPAC scores. The Student Record Books are available through the current Alpha Omega catalog; however, unlike the enclosed forms these books are not for duplication and should be purchased in sets of four to cover a full academic year.

WEEKLY LESSON PLANNER

			Week of:
Subject	Subject	Subject	Subject
Monday			
Subject	Subject	Subject	Subject
Tuesday			
Subject	Subject	Subject	Subject
Wednesday			
Subject	Subject	Subject	Subject
Thursday			
Subject	Subject	Subject	Subject
Friday			

WEEKLY LESSON PLANNER

Week of:

Subject	Subject	Subject	Subject
Monday			
Subject	Subject	Subject	Subject
Tuesday			
Subject	Subject	Subject	Subject
Wednesday			
Subject	Subject	Subject	Subject
Thursday			
Subject	Subject	Subject	Subject
Friday			

Student Name ______________________ Year ____________

Bible

LP #	Self Test Scores by Sections 1	2	3	4	5	Self Test Points	LIFEPAC Test	Oral Points	Report Points	Final Grade	Date
01											
02											
03											
04											
05											
06											
07											
08											
09											
10											

Language Arts

LP #	Self Test Scores by Sections 1	2	3	4	5	Self Test Points	LIFEPAC Test	Oral Points	Report Points	Final Grade	Date
01											
02											
03											
04											
05											
06											
07											
08											
09											
10											

Mathematics

LP #	Self Test Scores by Sections 1	2	3	4	5	Self Test Points.	LIFEPAC Test	Oral Points	Report Points	Final Grade	Date
01											
02											
03											
04											
05											
06											
07											
08											
09											
10											

Student Name ____________________ Year ________

Science

LP #	Self Test Scores by Sections 1	2	3	4	5	Self Test Points.	LIFEPAC Test	Oral Points	Report Points	Final Grade	Date
01											
02											
03											
04											
05											
06											
07											
08											
09											
10											

Social Studies

LP #	Self Test Scores by Sections 1	2	3	4	5	Self Test Points	LIFEPAC Test	Oral Points	Report Points	Final Grade	Date
01											
02											
03											
04											
05											
06											
07											
08											
09											
10											

Spelling/Electives

LP #	Self Test Scores by Sections 1	2	3	4	5	Self Test Points	LIFEPAC Test	Oral Points	Report Points	Final Grade	Date
01											
02											
03											
04											
05											
06											
07											
08											
09											
10											

TEACHER

NOTES

INSTRUCTIONS FOR SCIENCE

The LIFEPAC curriculum from grades two through twelve is structured so that the daily instructional material is written directly into the LIFEPACs. The student is encouraged to read and follow this instructional material in order to develop independent study habits. The teacher should introduce the LIFEPAC to the student, set a required completion schedule, complete teacher checks, be available for questions regarding both content and procedures, administer and grade tests, and develop additional learning activities as desired. Teachers working with several students may schedule their time so that students are assigned to a quiet work activity when it is necessary to spend instructional time with one particular student.

The Teacher Notes section of the Teacher's Guide lists the required or suggested materials for the LIFEPACs and provides additional learning activities for the students. The materials section refers only to LIFEPAC materials and does not include materials which may be needed for the additional activities. Additional learning activities provide a change from the daily school routine, encourage the student's interest in learning and may be used as a reward for good study habits.

If you have limited facilities and are not able to perform all the experiments contained in the LIFEPAC curriculum, the Science Project List for grades 3-12 may be a useful tool for you. This list prioritizes experiments into three categories: those essential to perform, those which should be performed as time and facilities permit, and those not essential for mastery of LIFEPACs. Of course, for complete understanding of concepts and student participation in the curriculum, all experiments should be performed whenever practical. Materials for the experiments are shown in Teacher Notes – Materials Needed.

Science Projects List

Key

(1) = Those essential to perform for basic understanding of scientific principles.
(2) = Those which should be performed as time permits.
(3) = Those not essential for mastery of LIFEPACs.

S = Equipment needed for home school or Christian school lab.
E = Explanation or demonstration by instructor may replace student of class lab work.
H = Suitable for homework or for home school students. (No lab equipment needed.)

Science 901

pp	4	(1)	H
	5	(2)	H
	6	(1)	S
	10	(1)	H or S

Science 902

pp	3	(1)	S
	7	(2)	S
	12	(1)	S
	16	(1)	S
	18	(1)	S
	24	(1)	H
	27	(2)	S
	30	(2)	H
	31	(1)	S

Science 903

pp	8	(1)	S
	35	(1)	S

Science 904

pp 10 field trip

Science 905-908

None

Science 909

pp			
	4	(1)	H
	8	(2)	H
	10	(2)	H
	11	(1)	H
	14	(1)	H
	23	(1)	S
	31	(3)	H

Science 910

none

Materials Needed for LIFEPAC

Required:
masking tape
two magnets marked with north and south poles one small piece of wood about the size of the magnets
clay—two colors ($\frac{1}{2}$ cup each)

Suggested:
pencil, block, ice cube
a balloon

Additional Learning Activities

Section I Structure of Matter

1. Help the student to research the size of atoms and compute the relative size of atoms of hydrogen, oxygen, and sulfur.
2. Discuss the three phases of matter. Have the students give examples for each phase. Write the examples on the board.
3. With your friends make flash cards containing the elements. Drill each other on the atomic number for each of the elements.
4. Make a chart showing the different phases of matter. Cut out pictures or draw pictures representing the different phases for your chart.

Section II Radioactivity

1. Demonstrate the use of a Geiger counter, if one is available.
2. With a friend plot intensity and distance on a graph similar to the one on Section III of LIFEPAC 901. Use the following numbers:

Intensity	Distance
6,572	1 cm
4,398	2 cm
3,221	4 cm
2,000	8 cm
1,582	16 cm
984	32 cm

3. Research the lives and discoveries of Marie and Pierre Curie. Write a one-page report on their discoveries and read it to the class.

Section III Atomic Nuclei

1 Explain the following formula: number of neutrons = atomic mass - atomic number. Work several problems on the board.
2. Have the students make a bulletin board that lists the seven particles and one ray that make up a complex nucleus.
3. With a friend draw and label a diagram of an atom.
4. Research Carl Anderson's life and discoveries and make an oral report to the class.

Section IV Nuclear Energy

1. Discuss fission and fusion. Write these words on the blackboard and ask the students to describe and compare the two.
2. Research the life and discoveries of Enrico Fermi. Write a one-page paper on the importance of Fermi's discoveries to nuclear science.
3. Make a chart showing the advantages and disadvantages of nuclear power.

Section V Nuclear Applications and Environmental Hazards

1. Visit a nuclear generating plant with a friend.
2. With a friend write down several ways in which atomic energy can be used for good purposes. Write several ways atomic power can be used for destructive purposes. Discuss your lists with the class. Do they agree with you? Ask your fellow students to add to both lists.

Materials Needed for LIFEPAC

Required:

junior baby-food jars, five small vials, five large jars, miscellaneous jars, one piece of wood, 2 cm x 14 cm x 20 cm; two pieces of wood, 2 cm x 14 cm x 15 cm; one dowel, 1 cm x 1 m; wire, 21 cm; one file; one piece of sandpaper; one small drill the size of the wire; two small half-pint milk cartons; four nails, 4 cm long one medium paper clip; white glue, box of paper clips, medium sized; five small objects; one large paper clip, several small paper clips, classroom equal-arm balance, one large paper clip, soap bubble kit, centimeter ruler, balance to measure mass, detergent, small beaker, graduated cylinder, catch container

Suggested:

100 ml graduated cylinder, 5 small rocks that must fit into the cylinder, hot plate, thermometer, small beaker of Ehrlenmeyer pyrex flask, ball of nonwater-soluble clay, paper clips, dishpan

Additional Learning Activities

Section I Volume

1. Copy liter patterns and have the students make construction paper cubes.
2. With a friend show that air takes up space by pushing a glass directly down into a bowl of water. What happens? Why?
3. With a classmate, calculate the volume of air in your classroom.
4. Calculate the volume of several books in your classroom. Use metric measurements.

Section II Mass

1. Show a filmstrip, video or movie on the metric system. Several should be available from your local library.
2. Lead a class discussion on why *mass* and not *weight* is the better scientific term to use. Ask the students what would happen to their *mass* and *weight* on the moon.
3. With a classmate plan and do an experiment that shows whether a certain volume of hot water weighs more or less than an equal volume of cold water.
4. Calculate your weight and the weight of a classmate in kilograms.

5. Find three objects, one that is square, one that is irregularly shaped, and one that is rectangular. Find the volume, mass, and density for each object.
6. Show that air has weight. Using a double-bar balance scale, balance an airless beach ball on one side and several small pieces of tinfoil or sand on the other. Blow up the beach ball and place it back on the scale. What happens?

Section III Density

1. Lead a class discussion on how density is affected by expansion and contraction.
2. With a classmate calculate the density of several items in the classroom by first determining their volume and mass.
3. If your teacher has a hydrometer, an instrument used for measuring the density of liquids, ask a classmate to help you measure the density of several liquids.
4. Do some research on submarines. Write a one to two page report explaining how they can travel below the surface and return to the top again.

Materials Needed for LIFEPAC

Required:	Suggested:
pencil, desk or small table, long sheet of paper, string, brick	
double-pan balance, set of metric masses, string, various rocks, beaker	

Additional Learning Activities

Section I Earth Structures

1. Organize a field trip to a local rock or mineralogy museum.
2. With a classmate gather several rock samples. Your teacher may have some you can use. Check each sample for color and weight. See if you can determine whether the rocks are igneous, sedimentary, or metamorphic.
3. With a classmate find the directions in a library science reference book for making a "volcano." Construct one to show your class.
4. In the library science resource books look up *minerals*. Write a one-page report on minerals and how they relate to rocks.

Section II Earth Changes

1. Demonstrate the effect of sedimentation by stirring dirt into a beaker full of water. Set the beaker aside. Lead a class discussion on how sedimentation takes place in the oceans and lakes.
2. With a classmate demonstrate the effects of mechanical weathering. Heat a piece of glass tubing over the flame of a Bunsen burner. Be sure to use forceps. When the tubing is very hot, plunge it into cold water. What happens?
3. Demonstrate another effect of mechanical weathering with a friend. Fill a glass bottle with water. Wrap the bottle with a towel and set it in the freezer overnight. Carefully check the bottle. What happened?
4. Pour a mixture of small gravel, sand, and clay into a jar of water. Stir vigorously and let stand. Check the jar later to see what material has settled to the bottom first. Were layers formed?

Section III Earth Movements

1. With a friend trace the continents of the world on a map. Then cut each continent out. See if you can fit the "puzzle" pieces together.
2. Write to the United States Geological Survey, Department of the Interior, Washington, D.C., 20244. Ask for a contour map of the area you live in.
3. Read a book on the drifting continents. Write a one-page report on the book. Be sure to include your scientific opinion of what is really happening to the continents. Make sure your opinion is reasonable based on your research.

Materials Needed for LIFEPAC

Required:

Suggested:
felt tip pen, one dozen plastic sandwich bags with ties, pen or pencil, two paper grocery bags, clipboard, twenty 3″ x 5″ cards

Additional Learning Activities

Section I An Observational Science

1. With a classmate visit the section on geology at your local science museum.
2. Do some research on how the earth's history has been divided into different periods. Make a chart showing the geologic time scale used in this country.
3. Write a two-page report on how the work with fossils done by Baron Cuvier and William Smith has helped geologists determine the age of different rock layers.

Section II Measuring Time

1. With your class use a world map or globe to find the location of the areas discussed in Section II.
2. With a friend make a sundial and use it to tell solar time. Most encyclopedias or general science books have directions for making sundials.
3. With a classmate study a map that has time zones. When it is four o'clock where you live what time is it in San Francisco? New York? Honolulu?
4. Write a two-page science fiction paper telling what you think historical geologists will find in the year 2037. What methods will they use to determine what life was like in 1996?
5. Make a wall chart illustrating the difference between relative time and absolute time.

Materials Needed for LIFEPAC

Required:

Suggested:
encyclopedia
dictionary

Additional Learning Activities

Section I Microorganisms

1. Obtain slides of one-celled organisms. Let your class view them through the microscope.
2. With a classmate tour a local hospital. Write a report on what you learned.
3. Volunteer to assist your school nurse for several days. Report to your class when you are finished.

Section II Bacterial Infections

1. Discuss any bacterial infections your students have had. Talk about the importance of cleanliness and sanitation in prevention of bacterial infection.
2. Write to health organizations asking for information on control of bacterial infections.
3. Research the disease, cholera. Write a one-page report on the effect it has had on history.

Section III Viral Infections

1. With a friend visit a local health department. Ask for information on viral infections.
2. With a friend make a poster on a viral infection. List the infection, symptoms, how it is spread, and how common it is in the United States.
3. Write the World Health Organization and ask for information on viral infections.
4. Research and write a one-page report on the treatment of rabies.

Section IV Protozoan, Rickettsial, and Fungal Infections

1. With a friend make a chart on the different types of typhus.
2. Grow your own mold or fungus.

Materials Needed for LIFEPAC

Required:	Suggested:
none	none

Additional Learning Activities

Section I. Disease Treatment

1. With a friend visit the local health department. Ask what is being done to prevent the spread of disease.
2. Research the importance of chemotherapy. Write a one-page report on the uses of chemotherapy. Read your report to the class.

Section II. Disease Prevention

1. Discuss with the class the importance of cleanliness and community sanitation in disease prevention. Have members of the class name all of our modern sanitary conveniences and list them on the board.
2. Take the students on a field trip to the local sewage plant. Stress the money, effort, planning, and technology that our society uses to keep us free from disease.
3. With a friend make a wall poster listing and illustrating seven factors of total health.
4. With a friend interview your classmates about their immunizations. Make a chart listing everyone's immunizations.

Section III. Community Agencies

1. Have the students write letters to the local, state, and federal health departments. Have them ask for information on disease treatment and disease prevention.
2. With a friend visit one of the volunteer groups in your community. Ask for information you can share with your classmates.
3. Write to the Food and Drug administration. Ask them to send you information on disease prevention.
4. Make a chart listing all of the volunteer agencies in your community. List the name, address, phone number, and purpose of the organization.

Materials Needed for LIFEPAC

Required:
none

Suggested:
none

Additional Learning Activities

Section I The Universe

1. Work numerous astronomical distance problems incorporating scientific notation to ensure student comprehension of superscript computations when large numbers are multiplied and divided.
2. Have the class make mathematical comparisons among the planets Mars, Venus, and Earth with respect to their mass, density, temperature, and surface gravity.
3. With a friend identify the constellations at night. Sketch and label the different constellations.
4. With a friend visit a local planetarium.
5. Obtain a picture or schematic representation of the Milky Way Galaxy which shows where our solar system fits into its vast elliptical dimensions.
6. Research the Greek astronomers. Describe the difference between ancient Greek and present-day astronomy.

Section II Telescopes and Optics

1. Show the class refracting and reflecting telescopes and explain the difference. If a physics laboratory is available which has a prism experiment, refraction can be more readily explained than through the use of a diagram.
2. With a friend look through a telescope and chart a constellation.
3. Research the life of Isaac Newton. Report to the class on his importance to the science of astronomy.
4. Research the critical new developments in telescopes. Write a one-page paper on the new developments.

Section III Space Exploration

1. Lead a class discussion concerning the estimate of statistical chance for the existence of extraterrestrial intelligent life.
2. Use the delphi-technique to attempt to get student input into technology breakthroughs required for interstellar travel.
3. Write a one-page report on what kind of life might exist on the other planets.

Materials Needed for LIFEPAC

Required:

Suggested:
almanac

Additional Learning Activities

Section I History of Oceanography

1. Research and make a large chart listing all of the important discoveries and innovations in oceanography.
2. Research and write a one-page report on the effect of the moon on tidal reaction.

Section II Geology of the Ocean

1. Have class construct a plaster of Paris model representing the major typographical features of the ocean floor, including such things as the mid-ocean ridges, rifts, deep trenches, seamounts, and continental shelves, all made to a reasonably accurate scale. Once the model is completed, it can be an excellent training aid to assist students in their understanding of the theories of sea-floor spreading and global tectonics.
2. With your friends write to the following organizations asking for charts, maps, and information on the geology of the ocean. You can find complete addresses at your local library: Naval Oceanographic Office (Suitland, MD), U.S. Coast Guard and Smithsonian Institution (Washington, D.C.), and the National Ocean Survey (Rockville, MD). Also, there are a number of educational and nonprofit institutes— Scripps Institution of Oceanography (La Jolla, CA) and Woods Hole Oceanographic Institution (Woods Hole, MA); Texas A&M (College Station, TX); Oregon State University (Corvallis, OR); University of Rhode Island (Kingston, RI); University of Washington (Seattle, WA); Johns Hopkins (Baltimore, MD); and the University of Miami (Miami, FL).
3. With a friend make a chart showing the ocean currents.
4. Write a one-page report explaining the effect of sea-floor spreading in producing oceanic trenches.
5. Research the life and discoveries of V.W. Ekman. Explain the Ekman Spiral in your report.

Section III Biological, Chemical, and Physical Properties

1. Have the class obtain data available in various world almanacs concerning world production and consumption of petroleum. Also, have them obtain clippings of recent magazine and newspaper stories of noteworthy events which bear on the political, economic, and social considerations relative to the oil-energy situation. Encourage the class to analyze the cause and effect relationship regarding decisions by

countries with regard to an energy crisis and how the science of oceanography may be the answer to an ever increasing problem.

2. Assign a student the task of making a full report of fish meal production; another one on its role in the food chain process which occurs off the coast of Peru; and a third on its value as a source of protein compared to other marine life.
3. Research the life of Jacques Cousteau. Write a paper on his importance to the science of oceanography.
4. With a friend make a poster of the carbon dioxide cycle.
5. Write a one-page paper explaining what causes ocean waves.

Materials Needed for LIFEPAC

Required:
screw-top jar lid, paper toweling, Plastic sandwich bag, bird seed, rubber band or wire tie, water

Suggested:
Jar with a lid, water, small amount of soil
Flat cake pan, sand (or soil), assorted small rocks, water
three clear glass containers of various sizes, medicine dropper, food coloring, water, watch with second hand
labels from various food products such as soups, fish, peanut butter, cereals, jellies, dairy products, coffee, butter, soy beans, beef
Plastic tray or metal pie pan lined with plastic wrap, sulfuric acid, wooden pencil, small piece of metal, small piece of plastic, paper napkin, small pieces of some type of meat product or dog food, water
Sunday edition of the newspaper, yardstick

Additional Learning Activities

Section I People and Their Land

1. Lead a class discussion on what your community or school could do to help the environment.
2. Organize several classmates into a clean-up committee. Spend time cleaning up your school, church, or other community area.
3. Make a list for several days of all the waste you see. For each item on your list, give a suggestion of how such waste could be avoided.
4. Look up one of the undeveloped nations in the library. (For example: India, several South American and African nations.) How is your choice affected by a growing population and poor agriculture? Write a one-page report on the nation.

Section II People and Their Work Environment

1. With a friend tour a local industry or factory. Check for cleanliness and lack of pollution. Write a one-page report on what you learned.
2. Draw a picture of what you think some form of transportation (car, bus, or train) will look like in the year 2010.
3. Listen to the weather report every day for a week to hear the "air" report. Many weather stations give the ozone, carbon monoxide, and pollution readings.

Section III People and Their New Frontiers

1. Discuss with your class why Christians need not fear the future and how Christians can help make the world better for all men. Reference Romans 8:35-39.
2. Paint a mural. Show a scene from the future in outer or inner space. Hang the mural in your classroom.
3. Write a one-page paper about your concerns for the future. How does Jesus help make that future more secure for you?
4. Imagine you are living on a future space station or in an underwater city. Write a letter to a friend living in an underwater city or space station.

Materials Needed for LIFEPAC

Required:
none

Suggested:
none

Additional Learning Activities

Section I Practical Uses of Length, Volume, Mass, and Weight

1. Discuss the reasons why the United States should become a metric nation.
2. Gather pamphlets and news articles that have metric units used in the articles. Create a metric reading center.
3. Visit a grocery or hardware store and make a list of items with metric labels.
4. Measure the height of every class member using a meter stick.
5. Make a poster featuring labels from food cans and boxes that are given in metric units.

Section II Practical Health

1. Go through a catalogue of camping gear and choose the proper equipment for a safe backpacking trip. Make a poster of your findings.
2. Plan and eat a safe lunch that could be carried on an all-day hike in summer.
3. Make a poster showing the activities of the World Health Organization.

Section III Practical Geology

1. Use a map of your area, and point out areas of geological interest. Ask students to tell about visits they may have made to the areas.
2. Cover a large balloon with papier mache and make a globe.
3. Clip newspaper or magazine articles dealing with earthquakes. Plot the areas on a map.

Section IV Practical Problems

1. Discuss the pros and cons of nuclear energy.
2. Discuss your city or county garbage and domestic sewage problems.
3. Keep track of all of the plastic thrown away by your family and school for one week. Determine the average per person. Multiply that amount by the number of persons in your state or province. List ways to conserve plastic products.
4. Make a poster giving the pros and cons of hydroelectric power.
5. Try heating water by sunlight in a variety of containers of different shapes and colors. Determine which type of container is best and why.
6. Research the nuclear accidents at Three Mile Island or Chernobyl. Write a paper reporting on either of the incidents.

ALTERNATE

TESTS & KEYS

Reproducible Tests
for use with the Science 900
Teacher's Guide

Name ________________________

Answer *true* or *false* (each answer, 1 point).

1. _______ A solid has definite shape, size, and mass.
2. _______ Neutrons are found in the atomic nucleus.
3. _______ Electrons and positrons are the same.
4. _______ Isotopes all have the same atomic mass.
5. _______ Gamma rays are not affected by a magnet.
6. _______ Neutrinos have been examined by scientists under microscopes.
7. _______ Enrico Fermi received a Nobel Prize for identifying new elements and discovering nuclear reactions.
8. _______ Cadmium is used to make control rods in fission reactors.
9. _______ Fission involves producing heavier elements and energy.
10. _______ Fossil fuels are plentiful today, and nuclear generating plants are not needed.

Match these items (each answer, 2 points).

11. _______ neutral particle
12. _______ uranium
13. _______ negative particle
14. _______ CO_2
15. _______ bent toward South Pole

a. electron
b. proton
c. neutron
d. element
e. compound
f. alpha particle

Write the letter for the correct answer on each line (each answer, 2 points).

16. The least dense form of matter is _______ .
    a. gas b. liquid c. solid
17. Radium was discovered by _______ .
    a. Fermi b. Becquerel c. Madame Curie
18. The intensity of a radioactive sample is measured by _______ .
    a. a cloud chamber b. a Geiger counter c. an X ray
19 $^{226}_{88}$Ra has _________ neutrons.
    a. 314 b. 138 c. 88
20. An atom of the element nitrogen, whose atomic number is 7, has __________ electrons in its outer shell.
    a. 7 b. 14 c. 5

21. In a nuclear reactor the fuel contains a fissionable material which _____ .
    a. provides unstable nuclei
    b. cools the heat
    c. moderates
22. Energy use per capita in the United States is increasing ______ .
    a. slower than the population
    b. the same as population growth
    c. more rapidly than population growth
23. The symbol for sodium is ______ .
    a. S   b. So   c. Na
24. The smallest of the three major particles of the atom is the ______ .
    a. electron   b. neutron   c. proton
25. The strongest in penetration of the three types of particles is ______ .
    a. alpha   b. beta   c. gamma

Complete these statements (each answer, 3 points).

26. The element fermium was named after ____________________ .
27. Marie Curie named ____________________ for the country of her birth.
28. The amount of material which, when brought together, would react spontaneously, is called ____________________ .
29. A spontaneous reaction that continues to feed itself and keep going is a ____________________ .
30. The unit of measurement of radiation biological material absorbs is the ____________________ .
31. Radioactive wastes are stored in abandoned ________________ mines.
32. The nucleus of the atom is made up of the a. ____________________ , and b. ____________________ , and the c. ____________________ orbits around the nucleus.
33. Stars are examples of a. ____________________ reactors; a nuclear reaction producing lighter elements and energy is b. ________________.
34. A particle like an electron with a positive charge is a ________________ .
35. Members of each element that have differing atomic masses are ______.
36. The scientist who determined that uranium gives off rays was ____________________ .
37. Three phases of matter are a. ______________ , b.____________ , and c. ____________________ .

73/91

Date ____________________
Score ____________________

Name ____________________

Write the letter for the correct answer on each line (each answer, 2 points).

1. A cube with a volume of 1,000 cm^3 is called 1 _______.
   a. quart — c. milliliter
   b. liter — d. gram
2. The international standard of mass is a kilogram of platinum kept in _______.
   a. Paris — c. Rome
   b. Washington — d. New York
3. The mass of 1.00 L of water is a _______.
   a. pound — c. kilogram
   b. milligram — d. gill
4. Archimedes is known for his discovery of _______.
   a. mass — c. gravity
   b. weight — d. density
5. An accepted reference unite that is mutually agreed upon as the basis of measurement for that characteristic is _______.
   a. accuracy — c. standard
   b. precise — d. metric

Answer *true* or *false* (each answer, 1 point).

6. _______ A gram is one-thousandth of kilogram.
7. _______ The location of an object may cause its weight to vary.
8. _______ An equal-arm balance is a recent development in measurement.
9. _______ Mass and weight mean the same thing.
10. _______ Volume determined by the displacement method is an indirect method of measurement.
11. _______ Mass divided by volume is density.

Complete these statements (each answer, 3 points).

12. The apparent loss of weight by a substance in a liquid is ____________.
13. The metric standard unit of mass is the ____________________.
14. Irregularly shaped solids are best measured by the _______________ (direct, indirect) method.
15. Newton's Law of ____________________ relates to the force of attraction between any two objects.
16. Grams per cubic centimeter express ____________________________.

Define these terms (each definition, 3 points).

17. weight ____
____
18. standard ____
____
19. accuracy ____
____
20. specific gravity ____
____

Calculate the answer (this answer, 8 points).

21. A rock has a mass of 250 gams and displaces 50 cm^3 of water. What is its specific gravity?

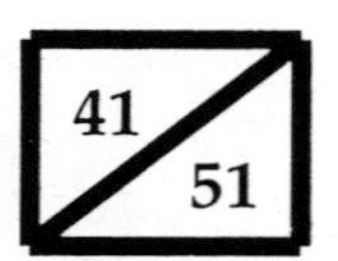

Date ____________________
Score ____________________

Name ________________________

Write the letter for the correct answer on each line (each answer, 2 points).

1. Granite is _______ .
   a. metamorphic b. sedimentary c. igneous

2. Sandstone is _______ .
   a. metamorphic b. sedimentary c. igneous

3. Basalt is _______ .
   a. metamorphic b. sedimentary c. igneous

4. Marble is _______ .
   a. metamorphic b. sedimentary c. igneous

5. Shale is _______ .
   a. metamorphic b. sedimentary c. igneous

6. Feldspar and mica converting to clay minerals is an example of _______ .
   a. chemical weathering c. erosion
   b. physical weathering d. ooze

7. Domed mountains are pushed upward by masses of _________ .
   a. lava c. gases
   b. magma d. sedimentation

8. Volcanic debris does *not* include _________ .
   a. lava c. tuff
   b. magma d. cinder

9. Gases in a volcano are primarily _______ .
   a. water vapor c. sulfur dioxide
   b. carbon dioxide d. nitrogen

10. Steep-sided glaciated valleys are called _______ .
   a. domes c. graded beds
   b. basins d. fjords

Complete these sentences (each answer, 3 points).

11. To tell whether an igneous rock cooled slowly or rapidly, check the size of the ____________________.
12. Magma that penetrates buried rocks, but does not reach the surface, is called ____________________ rock.
13. Magma that erupts from a volcano is called ____________________.
14. Bends in a river crossing a lowland are called ____________________.
15. The mass of a unit volume of material is called ____________________.

Answer *true* or *false* (each answer, 1 point).

16. _______ Sir Isaac Newton reasoned that the earth bulges slightly at the center.
17. _______ The three main categories of rock are igneous, sedimentary, and metamorphic.
18. _______ Siltstone is an example of a metamorphic rock.
19. _______ The middle part of the earth's core is liquid.
20. _______ Earthquakes are measured by a seismograph.
21. _______ Volcanoes that have fast flowing liquids will be tall and thin.
22. _______ Regolith is partly weathered rock.
23. _______ A branch of a river is called a distributary.
24. _______ The density of an object compared to the density of air is called specific gravity.
25. _______ Glaciers that travel between ridges are called valley glaciers.
26. _______ Lava may be called ashes, cinders, or bombs.
27. _______ Major mountain types include fold, fault, dome, and erosional remnants.
28. _______ Mechanical weathering always involves a chemical change.
29. _______ A lake that evaporates is called a playa.
30. _______ A teardrop-shaped moraine is called a drumlin.

Match these items (each answer, 2 points).

31.	_______	designer of the universe	a. weathering
32.	_______	first to calculate circumference of the earth	b. God
33.	_______	rocks changed by heat and pressure	c. seamount
34.	_______	Second layer of the earth	d. old river
35.	_______	former volcanoes now under the sea	e. moraine
36.	_______	flat-topped hill	f. Eratosthenes
37.	_______	chemical and mechanical process that breaks up rocks	g. anticline
38.	_______	crosses a nearly flat flood plain	h. mesa
39.	_______	snow falling down a mountain side	i. metamorphic
40.	_______	top of a fold	j. avalanche
			k. mantle

Complete these activities (each answer, 5 points).

41. Explain the difference between sedimentary and igneous rocks.

____

____

____

42. Explain the difference between mechanical weathering and chemical weathering.

____

____

____

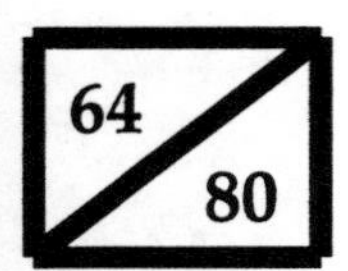

Date ____________________

Score ____________________

Name ____________________

Match these items (each answer, 2 points).

1. ________ turned to stone
2. ________ having a backbone
3. ________ the order in which rocks are placed one above the other
4. ________ science of fossils
5. ________ fragments, rubbish
6. ________ formed at the earth's surface
7. ________ in relation to others
8. ________ surface of erosion
9. ________ found in the sea
10. ________ gravel, sand, mud

a. relative
b. debris
c. unconformity
d. detrital sediments
e. marine
f. orogenic
g. vertebrate
h. sedimentary rocks
i. petrified
j. paleontology
k. superposition

Complete these sentences (each answer, 3 points).

11. Fossils that do not reveal the body form are called ______________.
12. Precipitation of a binding material around grains in rocks is called ______________.
13. The study of plants through the study of fossils is called ______________.
14. Orogeny is the cause of ______________.
15. A climatic calendar for the United States Southwest has been compiled from information gained from ______________.
16. Organisms without backbones are called ______________.
17. The group of scientists who believed that crustal rock arrived at the surface of the earth in molten state was called ______________.
18. Physics, chemistry, and biology are called ______________ sciences.
19. The parts of an organism that decay first are the ______________ parts.
20. The idea that younger formations overlie older formations is the law of ______________.

Complete these activities (each answer, 5 points).

21. Explain the difference between the Plutonists and the Neptunists.

    ____

    ____

    ____

22. State several reasons why sedimentary rocks are of interest to historical geologists. ____

    ____

    ____

    ____

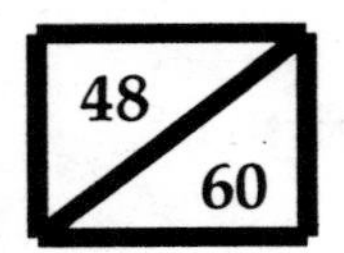

Date ____________________

Score ____________________

Name ____________________

Match these items (each answer, 2 points).

1. _______ disease germs
2. _______ Louis Pasteur
3. _______ bacteria
4. _______ protozoans
5. _______ fungi
6. _______ pathogenic
7. _______ tetanus
8. _______ dysentery
9. _______ hemorrhage
10. _______ symptom

a. one-celled organism that reproduces by fission
b. lockjaw
c. microorganisms
d. acute infection of the large intestine
e. founder of modern bacteriology
f. single-celled animals
g. mushrooms, mold, yeast
h. sign, indication
i. influenza
j. discharge of blood
k. producing disease

Answer *true* or *false* (each answer, 1 point).

11. _______ Children are usually born free of microorganisms.
12. _______ A toxin is a poison.
13. _______ We are still in the germ theory era.
14. _______ Some varieties of mold are used as antibiotics.
15. _______ Bacteria are classified as animals.
16. _______ A bodily infection is located in one area.
17. _______ A germ's worst enemy is cleanliness.
18. _______ Cholera is caused by a bacillus found in polluted drinking water.
19. _______ Ptomaine poisoning is food poisoning.
20. _______ Many working persons are required to have X rays to check for scarlet fever.
21. _______ Whooping cough is extremely contagious.
22. _______ Hippocrates described lockjaw 400 years before the birth of Christ.
23. _______ German measles is more severe than measles.
24. _______ Jonas Salk and Albert Sabin developed vaccines that are effective against smallpox.
25. _______ No cause for the common cold has been identified.

Write the letter for the correct answer on each line (each answer, 2 points).

26. Which of the following items is not a form of bacteria?_______
    a. bacillus
    b. coccus
    c. fungi
    d. spirochette
27. A cold without secondary infection usually lasts less than _______.
    a. two weeks
    b. one week
    c. three weeks
    d. five days
28. Yellow fever is spread by the bite of an infected _______.
    a. fly
    b. mosquito
    c. bee
    d. flea
29. Ringworm is caused by _______.
    a. rats
    b. mold
    c. ticks
    d. flies
30. Athlete's foot is actually _______.
    a. typhus
    b. influenza
    c. plague
    d. ringworm
31. Rickettsiae grow only in _______ cells.
    a. living
    b. dead
    c. infected
    d. young
32. Hepatitis is an infection of the _______.
    a. ear
    b. throat
    c. lungs
    d. liver
33. Which of the following diseases is not a typical childhood disease?______
    a. rabies
    b. measles
    c. chicken pox
    d. mumps

Complete these activities (each answer, 3 points).

34. List three ways disease may be spread.
    a. ________________________________
    b. ________________________________
    c. ________________________________
35. List three symptoms of a bodily infection.
    a. ________________________________
    b. ________________________________
    c. ________________________________
36. List three respiratory infections.
    a. ________________________________
    b. ________________________________
    c. ________________________________

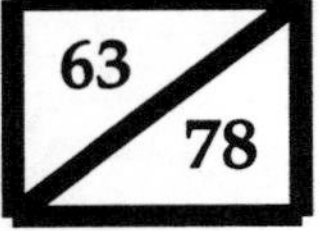

Date ________________

Score ________________

Name ____________________________

Complete these statements (each answer, 3 points).

1. Immunity is resistance to ____.
2. Hereditary conditions cause certain families to be more ______________ to one type of disease than another.
3. Epithelial cells of the openings of the body and their ________________ are a part of the body's defense mechanism.
4. Moving cells in the blood and lymph are called leukocytes or ________ ____.
5. An acid environment for microorganisms and protection for the body is provided by the____________________________________.
6. Interferon is a substance that interferes with the multiplication and growth of____.
7. An important body defense mechanism is elevated temperature or ________________.

Write the letter for the correct answer on each line (each answer, 2 points).

8. Antibiotics are used to help fight infection of______.
   a. fungus c. viruses
   b. bacteria d. fractures
9. An anti-infective suitable for treating malaria is______.
   a. salt c. iodine
   b. camphor d. quinine
10. Pneumonia is a bacteria infection caused by the pneumococcus bacteria and is usually treated with______.
   a. aspirin c. antiseptics
   b. formaldehyde d. antibiotics

Answer *true* or *false* (each answer, 1 point).

11. ______ Health education aids the local community in promoting better health.
12. ______ The county health department is the local health organization that has as its activities education, disease control, maternal and child care, and vital statistics.
13. ______ Volunteer health organization aid in the control of infectious disease.
14. ______ The primary organization for public health is the Food and Drug Administration.
15. ______ The Food and Drug Administration investigates all new drugs, adverse effects of drugs, additives, and foods.

Complete this activity (each answer, 3 points).

16. List five technological advances (such as the microscope invention) that were necessary to the advancement of medical progress.
    a. ______
    b. ______
    c. ______
    d. ______
    e. ______

Write the name of the scientist who did each of the following things (each answer, 3 points).

17 Discovered penicillin ______
18. Isolated antitoxin for diphtheria ______
19. Discovered cholera germ and established scientific study of bacteria ______
20. Invented rabies treatment and present milk-processing procedures ______
21. Began smallpox inoculations ______
22. Discovered moving cells called phagocytes ______
23. Produced oral polio vaccine used today ______
24. Founded chemotherapy–first chemical medicine ______

57/71

Date ______
Score ______

Science 907 Alternate Test

Name ____________________

Complete these statements (each answer, 3 points).

1. When we employ the power of base 10 to write very large or very small numbers, we are using ____________________.
2. The moons which orbit their planets are called ____________________.
3. In increasing order of the mean distance from the sun, the major planets are
   a. __________, b. __________, c. __________,
   d. __________, e. __________, f. __________,
   g. __________, h. __________, and i. __________.
4. The distance of 93,000,000 miles is called a(n)__________ .
5 A light-year measures____________________ miles.
6. The star which has an apparent magnitude of +0.8 is __________.
7. An optical telescope used to pick up the faintest visible stars is known as a ____________________ telescope.
8. The magnifying power of a telescope is computed by dividing the focal length of its objective by ____________________ .
9. In the Newtonian Reflector, light rays are bent at ____________________ before the image is formed.
10. Mars was explored by unmanned spacecraft called ____________________.
11. Skylab 4 placed astronauts into weightlessness for a period of ______ days.
12. The radio telescope at ____________________ measure 1,000 feet along its dish diameter.
13. Pioneer 10 and 11 made several passes around the planet __________ .
14. When we speak of life possibly outside of Earth, we use the term ____________________ .
15. Apollo 11 made a ____________________ landing.
16. Only the ____________________ has successfully landed men on the moon.
17. In order to undertake interstellar travel, we would need spacecraft with speeds near ____________________.

Match these items (each answer, 2 points).

18. ________ parallax
19. ________ recombines colors
20. ________ Chester Hall
21. ________ James Gregory
22. ________ Galileo's telescope
23. ________ 3-inch, f/8
24. ________ colored halo effect

a. theory behind achromatic refractors
b. suggested reflecting telescope
c. technique for measuring star distance
d. flint glass
e. 24-inch focal length
f. discovered gravity
g. 30X
h. chromatic aberration

Complete this activity (this answer, 5 points).

25. Calculate the degree of magnification for a 2″ and 5″ lens.

Complete this table (this answer, 3 points).

	Celestial Body	Rotation Period (days)
26.	Earth	________________
27.	Sun	________________
28.	Mars	________________
29.	Venus	________________
30.	Pluto	________________

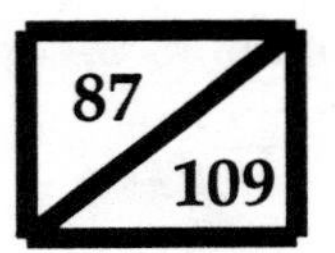

Date ________________

Score ________________

Name ________________________

Match these items (each answer, 2 points).

1. ______ Scott
2. ______ laser studies
3. ______ continental slopes
4. ______ diving saucer
5. ______ echo sounding
6. ______ East Indian Rift
7. ______ sedimentation process
8. ______ Woods Hole
9. ______ marine life extracted
10. ______ floating net-rafts
11. ______ underwater pump and electric lights
12. ______ upwelling
13. ______ 5°F increase world-wide
14. ______ krypton
15. ______ higher degree of salinity

a. Conshelf Two experiment
b. estimated to have occurred 20 million years ago
c. releases 15,000 bottles a year for current measurement
d. about 2/3 for human consumption
e. largely used by Japanese
f. produces about 99 per cent of fish available
g. biology of Antarctic whales
h. enough to raise ocean level by 250 feet
i. drop off 100-500 feet per mile
j. Ice Age
k. Cromwell Current
l. inert gas
m. echogram
n. lowers freezing point of water
o. Russian development for fish extraction

Complete these statement (each answer, points).

16. The most critical skill needed by mariners sailing across the sea was ______________________________.
17. A device used for attracting fish by use of sound reproduction is a ______________________________.
18. Sir Isaac Newton's law of gravitation provided a mathematical basis for ______________________________.
19. Whales are classified as____________________consumers in the ocean food web.
20. Laterally moving convection currents are thought to be a major reason for creating______________________________.

Write the letter for the correct answer on each line (each answer, 2 points).

21. The French tidal-power station produces______.
    a. 35,000 horsepower   b. 500 million kwh   c. tsunami
22. It is estimated that the world's petroleum reserves amount to______.
    a. 700 billion barrels   b. unlimited amounts   c. 1.2 trillion barrels
23. Confirmation of the ocean's deepest trench was achieved by the______.
    a. HMSM Challenger   b. Glomar Challenger   c. Trieste

24. The United States ranks fourth among fish-catch nations, but its fish consumption is _____________ of the world's average.
    a. double b. triple c. quadruple
25. During the massive turbidity current off Newfoundland, transatlantic cables were found snapped as much as________ away.
    a. 100 miles b. 50 miles c. 300 miles
26. The main sponsor of the Conshelf Two experiment was_____.
    a. Ecole Polytechnique
    b. French national petroleum office
    c. Woods Hole

46/57

Date ______________________

Score ______________________

Name ______________________

Match these items (each answer, 2 points).

1. _______ relationship of living things to their environments
2. _______ condition of becoming a city
3. _______ to lose
4. _______ disease or injury to plants
5. _______ science of drugs
6. _______ to extract or take out
7. _______ relating to tension or stress
8. _______ state of a species no longer living
9. _______ person who travels in space
10. _______ result of a lack of food
11. _______ conventional fuels
12. _______ consisting of land

a. blight
b. famine
c. astronaut
d. terrestrial
e. fossil fuels
f. dissipate
g. ecology
h. hydroelectric
i. tensile
j. extract
k. extinction
l. pharmacology
m. urbanization

Complete these activities (each answer, 3 points).

13. List three types of fossil fuels.
a. ______________________
b. ______________________
c. ______________________

14. List three problems of rapid population growth.
a. ______________________
b. ______________________
c. ______________________

15. List three items that can be effectively recycled.
a. ______________________
b. ______________________
c. ______________________

Answer true or false (each answer, 1 point).

16. _______ Nuclear energy can be generated by fission and fusion.
17. _______ Hydroelectric power involves the use of water.
18. _______ *Marine* refers to *things in space science.*
19. _______ The bathyscaph helped scientists learn about the ocean.
20. _______ Scientists are working hard to save extinct species.
21. _______ The biosphere is divided into land and water.
22. _______ Man lives in the biosphere.

Write the letter for the correct answer on each line (each answer, 2 points).

23. More than 70 per cent of the earth's surface is covered by________.
    a. pasture c. land forested
    b. water d. tree crop
24. A chemical used to prevent disease is called________.
    a. an antibiotic c. a hormone
    b. DDT d. a pesticide
25. A disease caused by a lack of protein is________.
    a. famine c. blight
    b. measles d. kwashiorkor
26. The population of the world is________.
    a. increasing c. stabilizing
    b. decreasing d. neutralizing
27. The process of splitting atoms is called________.
    a. nuclear c. fusion
    b. fission d. radiation
28. Hot springs power is also called________power.
    a. geothermal c. solar
    b. nuclear d. hydroelectric
29. Which of the following places is *not* a new frontier?________
    a. outer space c. the oceans
    b. the forest d. self exploration
30. Which of the following is not a natural power source?
    a. sun c. fossil fuels
    b. wind d. water
31. Before man can drink ocean water he needs to________it.
    a. extract c. purify
    b. irrigate d. desalinate
32. The energy released when atoms are split is called________.
    a. fission c. fusion
    b. radiation d. neutron

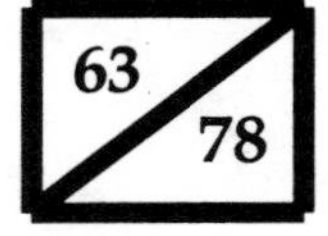

Date ______________________
Score ______________________

Name ____________________

Match these items (each answer, 2 points).

1. _______ skin
2. _______ food poisoning
3. _______ sun
4. _______ nucleus
5. _______ sedimentation
6. _______ trash
7. _______ mid-ocean
8. _______ smog
9. _______ thermal power
10. _______ vaccination

a. an injection for disease prevention
b. hot springs and geysers
c. the depositing of eroded material
d. dense central portion of an atom
e. major problem of developed nations
f. first line of body defense
g. thick haze
h. electron
i. area of volcanoes on sea floor
j. source of solar energy
k. common health problem on picnics

Answer *true* or *false* (each answer, 1 point).

11. _______ Parents are responsible for their children's immunizations.
12. _______ Geology is an experimental science.
13. _______ Fission is the splitting of a larger atom.
14. _______ Yellow fever and malaria are spread by mosquitoes.
15. _______ Hydroelectric power dams alter the appearance of scenic areas.
16. _______ Volcanoes cause slow changes in the landscape.
17. _______ Rich agricultural land is being destroyed by city growth.
18. _______ A proton is negatively charged.
19. _______ Hypothermia is caused by exposure.
20. _______ Ocean plants are major producers off oxygen.

Convert each item to the requested unit (each answer, 2 points).

21. 2 km = ______________ m
22. 2500 g = ______________ kg
23. 5.5 L = ______________ ml
24. 1.8 kg = ______________ g
25. 2,000ml = ______________ L

Complete this activity (each answer, 3 points).

26. List the three major units of the metric system.

    a. ______________________________
    b. ______________________________
    c. ______________________________

Define these words (each answer, 4 points).

27. weight

____

28. mass

____

Match these items (each answer, 2 points).

29. _______ fossil fuel
30. _______ ticks
31. _______ electron
32. _______ upbuilding
33. _______ fusion
34. _______ erosion
35. _______ plate tectonics
36. _______ earthquake
37. _______ potable
38. _______ neutron

a. washing away of soil
b. kilogram
c. neutral charge
d. safe to drink
e. theorizes that the continents were once one land mass
f. breaking and moving of rock
g. oil, gas, coal
h. negative charge
i. volcanoes, folds, faults, and sedimentation
j. joining of smaller atoms with release of energy
k. Rocky Mountain spotted fever

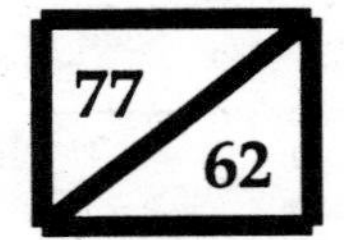

Date ____________________
Score ____________________

Notes

1. true
2. true
3. false
4. false
5. true
6. false
7. true
8. true
9. false
10. false
11. c
12. d
13. a
14. e
15. f
16. a
17. c
18. b
19. b
20. c
21. a
22. c
23. c
24. a
25. c
26. Enrico Fermi
27. polonium
28. critical mass
29. chain reaction
30. roentgen
31. salt
32. Any order (a. and b.):
    a. proton
    b. neutron
    c. electron
33. a. fusion
    b. fission
34. positron
35. isotopes
36. Becquerel
37. Any order:
    a. solid
    b. liquid
    c. gas

1. b
2. a
3. c
4. d
5. c
6. true
7. true
8. false
9. false
10. true
11. true
12. buoyancy
13. kilogram
14. indirect
15. gravitation
16. density
17. Weight is a measure of the pull between two objects.
18. Standard is that quantity against which others are measured.
19. Accuracy is closeness to a standard.
20. Specific gravity is the ratio of the density of a substance divided by the density of water.
21. Since, in the metric system, density and specific gravity are equal numerically (but without units) use $D = \frac{m}{v} = \frac{250g}{50cm^3} = 5\ g/cm^3$; Sp.G. = 5.

1. c
2. b
3. c
4. a
5. b
6. a
7. b
8. b
9. a
10. d
11. crystals
12. intrusive
13. lava
14. meanders
15. density
16. true
17. true
18. false
19. false
20. true
21. false
22. true
23. true
24. false
25. true
26. true
27. true
28. false
29. true
30. true
31. b
32. f
33. i
34. k
35. c
36. h
37. a
38. d
39. j
40. g
41. Example:
    Sedimentary rocks are formed in layers under the sea; igneous rocks are fire-formed.
42. Example:
    Mechanical weathering involves a physical change, and chemical weathering involves a chemical change.

1. i
2. g
3. k
4. j
5. b
6. h
7. a
8. c
9. e
10. d
11. trace fossils
12. cementation
13. paleobotany
14. crustal changes
15. tree rings
16. invertebrates
17. Plutonists
18. laboratory
19. fleshy
20. superposition
21. Example:
    Plutonists believed that crustal rock arrived at the surface of the earth in molten state. Neptunists believed that all crustal rock was precipitated from an ocean.
22. Example:
    Sedimentary rocks form at the earth's surface, are the burial ground for former life, preserve a record of life and environments, and represent the passage of time.

1. c
2. e
3. a
4. f
5. g
6. k
7. b
8. d
9. j
10. h
11. true
12. true
13. false
14. true
15. false
16. false
17. true
18. true
19. true
20. false
21. true
22. true
23. false
24. false
25. true
26. c
27. a
28. b
29. b
30. d
31. a
32. d
33. a
34. Examples; any order:
    a. coughing
    b. contaminated water
    c. insect bites
    or fecal contamination or contaminated food
35. Examples; any order:
    a. fever
    b. weakness
    c. loss of appetite
36. Examples; any order:
    a. pneumonia
    b. tuberculosis
    c. scarlet fever
    or whooping cough

1. disease
2. susceptible
3. secretions
4. white cells
5. skin
6. viruses
7. fever
8. b
9. d
10. d
11. true
12. true
13. false
14. false
15. true
16. Examples; any order:
    a. immunology
    b. X ray
    c. penicillin
    d. anaesthesia
    e. blood typing
17. Alexander Fleming
18. Emil Behring
19. Robert Koch
20. Louis Pasteur
21. Edward Jenner
22. Ilyn Mechnikov
23. Sabin or Salk
24. Paul Ehrlich

1. scientific notation
2. satellites
3. a. Mercury
   b. Venus
   c. Earth
   d. Mars
   e. Jupiter
   f. Saturn
   g. Uranus
   h. Neptune
   i. Pluto
4. Astronomical Unit
5. 6 trillion
6. Betelgeuse
7. reflecting
8. the focal length of its eyepiece
9. right angles
10. Viking I and II
11. 84 days
12. Arecibo
13. Jupiter
14. extraterrestrial
15. lunar
16. United States
17. the velocity of light
18. c
19. d
20. a
21. b
22. g
23. e
24. h
25. $\frac{\pi(2/2)^2}{\pi(5/2)^2} = \frac{1}{25/4} = \frac{4}{25}$
    Therefore, 4:25.
26. 1
27. 27
28. 1
29. -243
30. 6.4

1. g
2. k
3. i
4. a
5. m
6. b
7. j
8. c
9. d
10. e
11. o
12. f
13 h
14. l
15. n
16. navigation
17. recording
18. tidal theory
19. secondary
20. ocean trenches
21. b
22. a
23. c
24. a
25. c
26. b

1. g
2. m
3. f
4. a
5. l
6. j
7. i
8. k
9. c
10. b
11. e
12. d
13. Examples; any order:
    a. oil
    b. natural gas
    c. coal
14. Examples; any order:
    a. competition for necessities
    b. psychological harm
    c. lack of medical care
15. Examples; any order:
    a. glass bottles
    b. aluminum cans
    c. ash and cinders
16. true
17. true
18. false
19. true
20. false
21. false
22. true
23. b
24. a
25. d
26. a
27. b
28. a
29. b
30. c
31. d
32. b

1. f
2. k
3. j
4. d
5. c
6. e
7. i
8. g
9. b
10. a
11. true
12. false
13. true
14. true
15. true
16. false
17. true
18. false
19. true
20. true
21. 2,000m
22. 2.5 kg
23. 5,500 ml
24. 1,800 g
25. 2 L
26. Any order:
    a. gram
    b. meter
    c. liter
27. Example:
    Weight is the pull of gravity on an object.
28. Example:
    Mass is the amount of material in an object
29. g
30. k
31. h
32. i
33. j
34. a
35. e
36. f
37. d
38. c

LIFEPAC

ANSWER KEYS

SECTION ONE

1.1 Example:

It has shape, size and mass. It does not move around. It stays where you put it.

1.2 a.

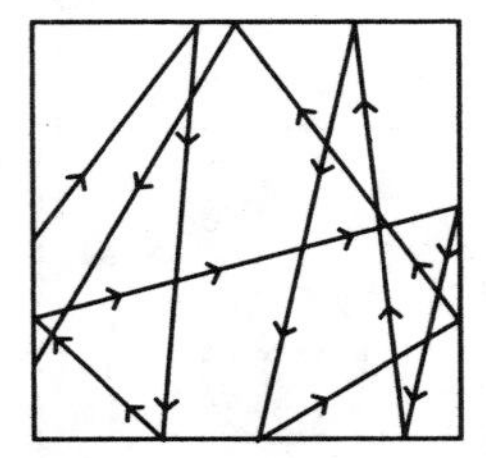

b. Moved in straight lines; had little freedom to move, turn, or change direction; motion very restricted.

1.3 a.

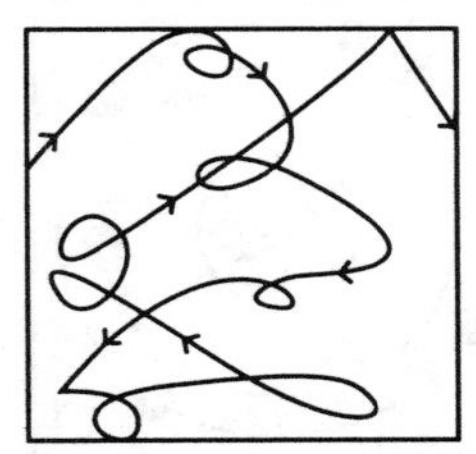

b. Can now move from place to place and turn (rotate) as I move; have more freedom, also move faster than as a solid.

1.4 Liquids must be restrained to remain in one spot. Liquids flow easily. Liquids take on the shape of the container. Liquids are soft. Liquids have mass. Liquids have a flat topped surface.

1.5 The molecules press against the sides, hitting the sides and pushing them back. The more gas, the more push.

1.6 The balloon flies around because the gas escapes, lowering the pressure on the outlet side which forces the balloon forward.

1.7 Gases occupy the entire container and must be totally covered or they will escape. The distance between particles is great because even when packed into a balloon they can't be seen. The density is very low. There is mass (because gas is matter) but the amount and volume is less than that of solids or liquids.

1.8 I have freedom of movement in all directions without restrictions. I collide infrequently with other particles. My speed is greatly increased.

1.9 parent check

1.10 repel

1.11 a. repel
b. repel
c. repel
d. repel
e. attract
f. attract

1.12 nothing

1.13 nothing

1.14 no

1.15 a. repel
b. repel
c. attract
d. neither
e. no
f. Because the n behaves like the wood and is not affected by the charged particles.

1.16 a. 1
b. 6
c. 7
d. 8
e. 12
f. 13
g. 16
h. 20
i. 26
j. 53

1.17 1
6
7
8
12
13
16
20
26
53

1.18 1
6
7
8
12
13
16
20
26
53

Note: Placement of electrons on a particular ring may be done in any location on that ring or shell.

1.19 a.

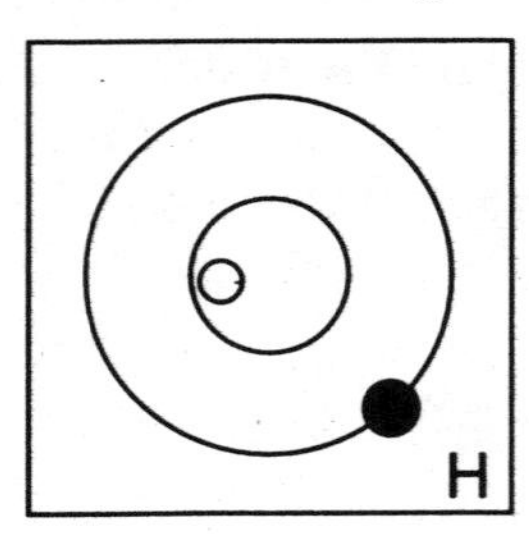

b.

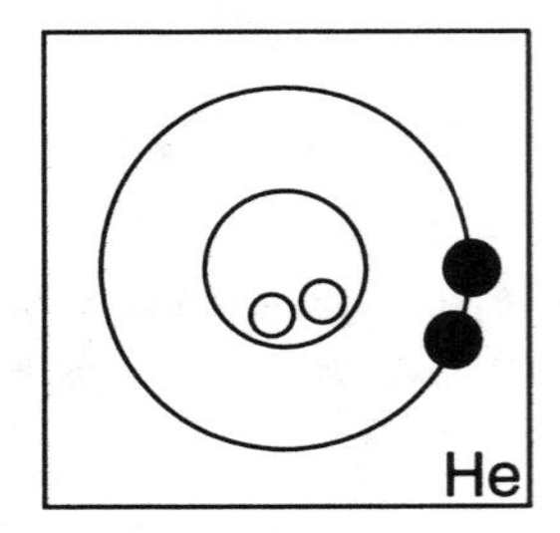

c.

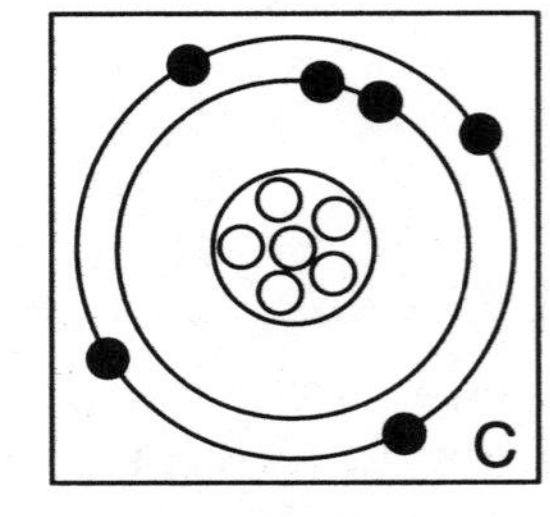

d.

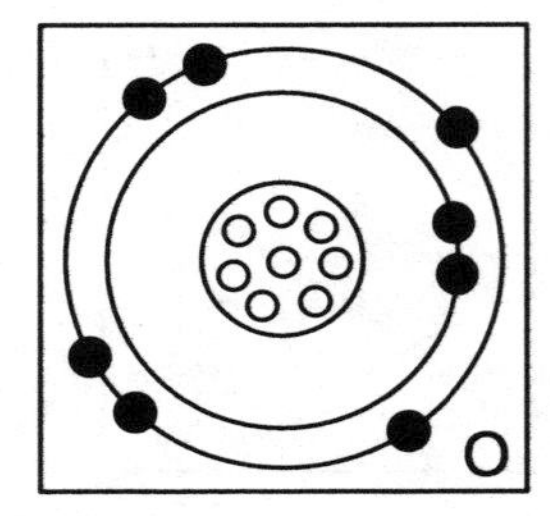

e.

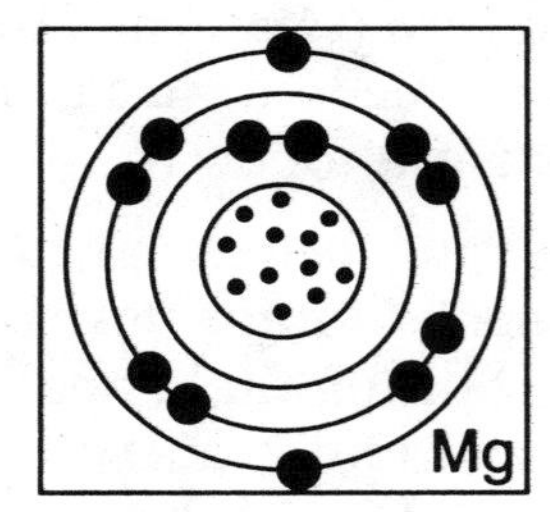

f.

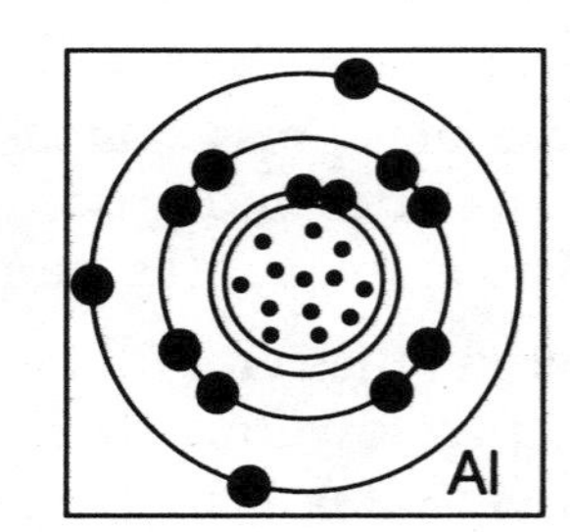

g. 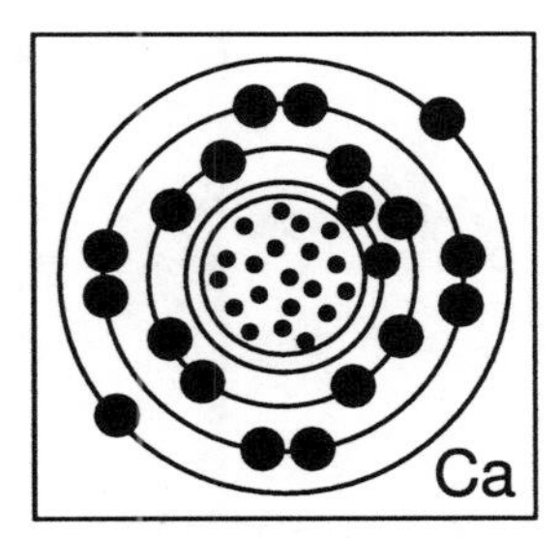

h. 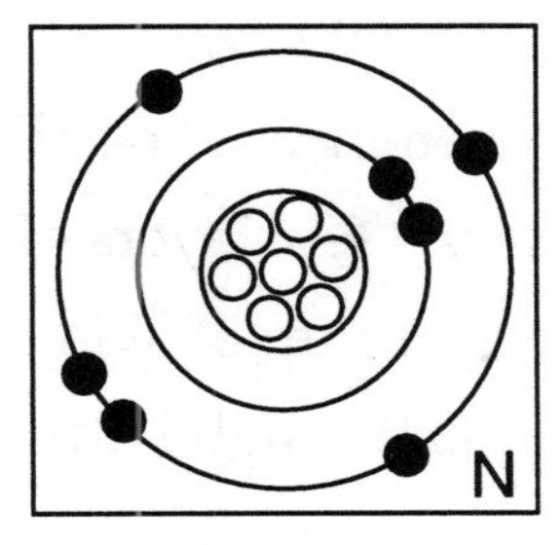

1.20 two thousand

1.21 Either order: (for a and b)

a. protons

b. neutrons

c. electrons

1.22 electrons

1.23 a. proton

b. electron

c. neutron

1.24 1

2

5

29

13

47

14

17

82

SECTION TWO

2.1 Either order:

a. photographic plates

b. the magnet

2.2 Any order:

a. Beta particles are bent sharply toward the North Pole.

b. Alpha particles are slightly bent to the South Pole.

c. Gamma rays are not affected.

2.3 Example:

Becquerel wrapped some uranium ore in papers and set it in a drawer. Unknowingly, he had set it on an undeveloped photographic plate. He discovered later that in the places where he had laid the plate, it developed as if it had been exposed to light.

2.4

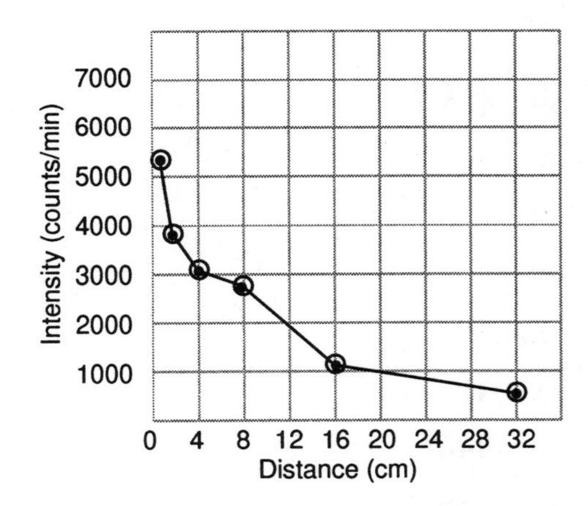

2.5 Decreases because the further you get from the source, the chance of being hit decreases.

2.6 The Wilson cloud chamber detects the presence of radioactive materials as well as speed and mass. Geiger counter measures the quantity (intensity) of radiation striking a certain area of space.

SECTION THREE

3.1
a. 1
b. 6
c. 7
d. 8
e. Na
f. Al
g. Ni
h. Fe

3.2
a. oxygen
b. atomic mass
c. number of protons (atomic number)

3.3
a. ${}^{1}_{1}H$
b. ${}^{12}_{6}C$
c. ${}^{35}_{17}C1$
d. ${}^{2}_{1}H$
e. ${}^{13}_{6}C$
f. ${}^{36}_{17}C1$

3.4
a. 2
b. 8
c. 7
d. 10
e. 12
f. 146
g. 138
h. 124
i. 0
j. 5

3.5
a. β
b. α
c. Υ
d. near speed of light, very fast
e. very fast (21,00 km/sec.)
f. bundle of energy at speed of light
g. stopped by a few layers of skin
h. stopped by 1 mm of skin
i. strong-pass through body

3.6 Any order:
a. electrons
b. neutrinos
c. mesons
d. protons
e. neutrons
f. positrons
g. alpha particles

3.7 gamma ray

3.8
a. 75
b. 77
c. 78
d. 80

3.9 One model has the "glue" of the nucleus, the buffer between the protons, being the neutrons and other particles of the nucleus. Gamma rays result when proton or neutron drops to a lower level in the nucleus.

3.10 When the protons and neutrons increase and the n:p ratio is greater than 1.5, the nucleus is not balanced and flies apart.

SECTION FOUR

4.1 1938 for identifying new elements and discovering nuclear reactions.

4.2 teacher check

4.3 The breaking of heavy, complex nuclei into smaller masses (atoms or particles).

4.4 ${}^{1}_{0}n$, ${}^{90}_{38}Sr$

4.5 A chain reaction is a nuclear reaction in which one of the products of nuclear decay initiates the decay of another atom. This can be controlled or stopped by inserting a substance that will absorb or buffer the decay products from the unstable nuclei.

4.6 Critical mass is a amount of mass of a substance that if brought together in one pile will self destruct by a spontaneous chain reaction.

4.7 helper check

4.8
a. Fuel rods provide the unstable nuclei.
b. Moderator slows the neutrons so they are correct energy for fission.
c. Control rods control the number of neutrons available to collide and fission the U-235.
d. Coolant is a liquid that absorbs the heat of a fission and heats the water for the steam turbine.
e. Shielding is concrete or lead that shields the environment from radiation.

4.9 teacher check

4.10 Fusion is the combining of light nuclei to form heavier atoms and release energy. Fission is a process of decay (tearing down) while fusion is a process of building. Much energy is produced by both reactions.

4.11 Our knowledge of the destructive power of nuclear energy makes it easier to comprhend the judgement of God upon the earth. We have concrete examples of the type of burning that can destroy trees, grass and mountains as referenced by the passage in Revelation.

SECTION FIVE

5.1

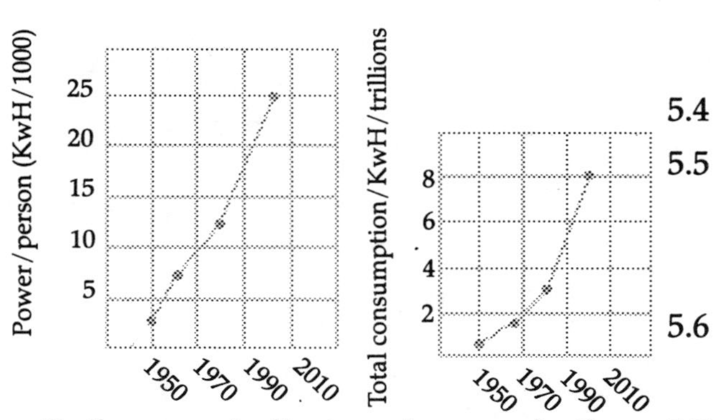

5.2 Both curves indicate an increase in the rate of consumption and the rate continues to be greater than linear.

5.3 Either order:
a. increasing population
b. increasing use of electricity

5.4 quadrupled

5.5 Either order:
a. use up oxygen
b. add heat and chemicals

5.6
a. waste (heat)
b. low levels of radioactivity

5.7 teacher check

Examples:
a. How did you feel?
b. Were you afraid?

a. How did you feel?
b. Were you afraid?
c. Did the treatments make you sick?
d. What did the doctors do?
e. How does this affect your relationship to God?
f. What does the treatment do to cancer?
g. How frequently are you treated?
h. What is your diet?

5.8 teacher check

5.9 Either order:
a. curie
b. roentgen

5.10 Materials are concentrated and made into solids, then placed in mines. Mines are safe because they're geologically stable, not connected to water sources, and water safe (water can't get in).

5.11 The effect heated water has on the original source when it is returned.

5.12 Either order:
a. irrigation of crops, keep oceans from freezing, and
b. prolongs shipping seasons, promotes growth in the oceans

5.13 helper check

SECTION ONE

1.1 Example data:
   a. 10
   b. 6
   c. 29
   d. 18
   e. 42

1.2 Example data:
   a. 9.5
   b. 6.5
   c. 29.5
   d. 17.5
   e. 42.0

1.3 Example data:
   a. 9.7
   b. 6.3
   c. 29.3
   d. 17.6
   e. 42.2

1.4 The smaller the division, the more accurately the volume can be measured.

1.5 Example:
   a. 0.1
   b. 0.0
   c. 0.2
   d. 0.2
   e. 0.4

1.6 Example:

For measuring very small volumes, even the $\frac{1}{10}$ divisions are not very accurate because number two had some water but not enough to be measured as $\frac{1}{10}$ of a unit.

1.7 no

1.8 Example:

We all worked separately and probably did not divide our jar exactly the same way. Therefore, the measurements will all be a little different.

1.9 no

1.10 It is impossible to do the dividing exactly the same way each time. Errors are bound to be present.

1.11 Example:
   1. Whether or not the same division methods were used.
   2. Care in measuring.
   3. Size of the divisions being made.

1.12 Divisions are to measure the volumes in between. You could have some volume but not enough to be more than 1/2 of the smallest unit, so by measurement you would have zero volume.

1.13 Example:

That quantity against which others are measured.

1.14 The accuracy decreases because copy is not exact.

1.15 1,000

1

.602

1.16 10,000

10

0.062

1.17 100

0.00001

6.200

1.18 10

0.1

0.0062

1.19 Sample data
- a 16.2, 19.4, 3.2
- b. 25.0, 29.2, 4.2
- c. 26.4, 39.2, 12.8
- d. 13.4, 26.2, 12.8
- e. 19.6, 32.8, 13.2

1.20 Answers will vary. Probably all will include some type of water displacement idea.

1.21 Answers will vary depending on plan in 1.20.

1.22 Air spaces and other factors may cause the solid to have an apparent size greater than its real size.

1.23 Gases are compressible and therefore the same air could occupy several different volumes depending upon pressure and/or temperature conditions.

SECTION TWO

2.1 no

2.2 To measure accurately, smaller units are needed.

2.3 Accuracy will increase.

2.4 Answers will vary.

2.5 Answers will vary.

2.6 The 1/10 unit measurements are much more accurate.

2.7 Example data:
- a. Dotty 4.6
- b. Jon 4.8
- c. Rex 4.5
- d. Irene 4.6
- e. Chuck 4.4

2.8 The balance does not reproduce exactly the same reading each time.

2.9
1. Care in measurement.
2. Friction of the axle.
3. Care in constructing the balance.
4. Sensitivity of the balance itself.
5. Wind or movement of the beam.

2.10 Answers will vary; might include:
1. Has upright support posts
2. Balances a rod on the posts
3. Pans on each end of the rod to hold object and mass units
4. Balance objects against standard mass units.

2.11 Accuracy = how close is the measurement to the real, actual value. Precision = how consistently you can measure an object or a value and get the same answer.

2.12 Standards are necessary to insure consistency of measurement from one time to another and from one place to another. Standards give uniformity to measurement.

2.13 Answers will vary.

2.14 Answers will vary.

2.15 Answers will vary.

2.16 The two values are close but not exactly the same. The classroom balance gives the most accurate value mass as it is the most accurate balance.

2.17
- a. .001
- b. .001
- c. .01
- d. .0001
- e. .0035
- f. .620
- g. 1000
- h. 1
- i. 0.1
- j. 100
- k. 3.5
- l. 1,000
- m. 1,000,000
- n. 10,000
- o. 100,000
- p. 620,000

2.18 The weight will be many times greater due to the greater mass of the sun.

2.19 Mass is a constant quality measuring the amount of "stuff" in an object while the weight is a variable measure of the attraction (pull) between two bodies.

SECTION THREE

3.1 Example data:
a. 3.0 cm
b. 4.2
c. 3.6
d. 2.8
e. 3.4

3.2 teacher check
Example data:
a. 14.0 cm^3
b. 38.5 cm^3
c. 24.3 cm^3
d. 11.5 cm^3
e. 20.4 cm^3

3.3 teacher check
Example data:
a. 0.0168 g
b. 0.04620 g
c. 0.02916 g
d. 0.01380 g
e. 0.02448 g

3.4 0.15 g/cm^3

3.5 0.07 g/cm^3

3.6 1.8×10^{-4} g/cm^3

3.7 8.4×10^{-5} g/cm^3

3.8 decreased

3.9 The amount of "stuff" cm^3 (density) decreases as a substance changes from a liquid to a gas.

3.10 As a substance evaporates, its density decreases.

3.11 1.2×10^{-3} g/cm^3

3.12 9×10^{-5} g/cm^3

3.13 The density has greatly decreased because the pressure of the air piled on top gets less as you get farther away from the surface of the earth. Since the pressure decreases, so does the density.

3.14 Sample data:
a. 25°C 0
b. 35°C 0.2 cm
c. 45°C 0.4 cm
d. 55°C 0.6 cm
e. 65°C 0.8 cm
f. 75°C 1.0 cm

3.15 A volume increases as the temperature increases.

3.16 increases

3.17 decreases

3.18 Answers will vary. Width, capacity, wall thickness

3.19 Answers will vary, but must explain Archimedes' Principle.

3.20 Sample data:
a. 15.2 g, 6.1 g, 2.5
b. 25.3 g, 5.2 g, 4.9
c. 9.8 g, 2.0 g, 4.9
d. 28.6 g, 9.2 g, 3.1
e. 12.4 g, 3.2 g, 3.9

3.21
1. Spilling overflow water
2. Inaccurate measurement of mass and volume

SECTION ONE

1.1 Eratosthenes
1.2 Ptolemy
1.3 sphere
1.4 During a lunar eclipse, the earth's shadow on the moon is circular.
1.5 The width was one-eighth of a minute of latitude.
1.6 true
1.7 true
1.8 true
1.9 igneous
1.10 lava
1.11 magma
1.12 Either order:
  a. tuff
  b. volcanic ash
1.13 b-c; either order:
  a. quartz
  b. feldspar
  c. mica
1.14 basalt
1.15 a. cooled from magma or lava
  b. laid in place by moving water, ice, or wind
  c. put under pressure, or under heat and pressure
1.16 a. sandstone
  b. conglomerate
  c. breccia
  d. siltstone or shale
1.17 a. slate or schist
  b. marble
  c. quartzite
1.18 true
1.19 false
1.20 true
1.21 true
1.22 false
1.23 false
1.24 c
1.25 a
1.26 c
1.27 b
1.28 hydrosphere
1.29 basalt
1.30 crust
1.31 mantle
1.32 asthenosphere
1.33 The gravitational pull of the earth is greater than the weight of surface rock.
1.34 Either order:
  a. earth's magnetic field
  b. meteorite composition
1.35 Either order:
  a. compression waves
  b. sideways shaking motions (vibrations)
1.36 Sideways vibrations do not move through liquids; sideways shock waves get lost at 2,900 kilometers.
1.37 a. The line is straight down the paper.
  b. The line is wavy and jagged rather than straight and smooth.
  c. The line has big notches (jagged marks) and these get smaller down to a straight even line.
1.38 true
1.39 true
1.40 false
1.41 false
1.42 false
1.43 c
1.44 b
1.45 Either order:
  a. silicon
  b. oxygen
1.46 size
1.47 pressure
1.48 intrusive
1.49 sill
1.50 joints
1.51 shield
1.52 Either order:
  a. silicon
  b. oxygen
1.53 Any order:
  a. potassium

b. aluminum
c. sodium
d. magnesium
e. calcium or iron
1.54 Slow cooling allows time for molecules to move and come in contact with other similar molecules.
1.55 teacher check
1.56 false
1.57 true
1.58 true
1.59 true
1.60 false
1.61 a
1.62 c
1.63 Lava flows flow out of a fissure. A volcano is the result of lava finding a small place where the ground is weak.
1.64 huge waves of ocean water
or dust in the sky
1.65 teacher check
1.66 false
1.67 true
1.68 false
1.69 scarp
1.70 magma
1.71 mesas
1.72 erosion
1.73 Either order:
a. along zones of weakness
b. over "hot spots"
1.74 Any order:
a. shields of lava flows
b. cinder cones
c. combinations of a and b
1.75 Any order:
a. volcanoes
b. folded mountains
c. fault-block
d. domes
e. erosional remnants

SECTION TWO

2.1 true
2.2 false
2.3 true
2.4 a
2.5 b
2.6 They can be turned into rock (stone).
2.7 Either order:
a. chemical
b. mechanical
2.8 Either order:
a. plant roots
b. decaying plants
2.9 exfoliation
2.10 talus
2.11 clay
2.12 Any order:
a, on the floor plain of a river
b. in a lake
c. at the mouth of a river
d. on a sandbar
e. in a sand dune
2.13 Any order:
a. wind
b. water
c. ice
2.14 Either order:
a. dissolves minerals
b. freezes and expands
2.15 topsoil - decayed vegetation
subsoil - weathered rock
regolith - partly weathered rock
bedrock - unweathered rock
2.16 true
2.17 false
2.18 false
2.19 true
2.20 false
2.21 true
2.22 true
2.23 true

2.24 c
2.25 a
2.26 d
2.27 d
2.28 a
2.29 c
2.30 d
2.31 b
2.32 a
2.33 headward erosion
2.34 alluvial fan
2.35 playa
2.36 flood plain
2.37 oxbow
2.38 old
2.39 desert
2.40 loess
2.41 plucking
2.42 striated
2.43 till
2.44 Any order:
 a. Desert sand is more uniform in grain size.
 b. Beach sand contains more silt.
 c. Beach sand may contain marine fossils.
2.45 A valley glacier is confined by ridges and flows downhill. A continental glacier covers areas hundreds of miles wide.
2.46 Any order:
 a. erratic boulders
 b. eskers
 c. moraines
 d. scratches (striations)
 e. drumlins
2.47 false
2.48 true
2.49 true
2.50 false
2.51 true
2.52 false
2.53 false
2.54 true
2.55 c
2.56 c
2.57 b
2.58 a
2.59 a
2.60 distributary
2.61 estuary
2.62 continental shelf
2.63 oolites
2.64 bacteria
2.65 ooze
2.66 reef
2.67 varves
2.68 water table
2.69 cavern
2.70 a. the rocks are broken into smaller pieces
 b. the rocks are worn smooth
2.71 Storms remove sand, leaving cobbles. Storms can destroy a beach.
2.72 Either order:
 a. nearby beach cliffs
 b. river - transported sand from hills and mountains
2.73 Nutrients (minerals) area available to grow algae which is eaten by fish and other sea life.
2.74 Rivers on land and ocean currents do not have enough energy to transport coarse material beyond the continental shelf.
2.75 lake -bog - solid ground (by infilling) or lake - river - (by destruction of the dam)
2.76 Any order:
 a. lake
 b. river
 c. spring

SECTION THREE

3.1 true
3.2 true
3.3 true
3.4 true
3.5 true
3.6 true
3.7 false
3.8 a
3.9 c
3.10 density
3.11 fjords
3.12 isostasy
3.13 under high pressure
3.14 evaporation of seawater
3.15 salt domes
3.16 density $= \frac{\text{mass}}{\text{volume}} = \frac{4 \text{ grams}}{2 \text{ cm}^3}$

$2 \frac{\text{grams}}{\text{cm}^3}$

3.17 2 (no units)
3.18 Plastic means that the rock will have a new shape or that it will flow due to heat and/or pressure.
3.19 A floating object is held up by a force equal to the weight (mass) of the substance displaced. For example, if a boat weighs 2,000 grams out of water and 1,000 in water, the boat in the water takes up the space of 1,000 grams of water.
3.20 The ice sheet is heavy enough to push the middle of the Antarctic continent down below sea level.
3.21 Either order:
  a. mining of salt
  b. as petroleum traps
3.22 c
3.23 d
3.24 geosynclines
3.25 isostasy
3.26 plateau
3.27 Either order:
  a. a former mountain range
  b. a more widespread source
3.28 b
3.29 a
3.30 false
3.31 true
3.32 true
3.33 true
3.34 Africa
3.35 Gondwanaland
3.36 rift valleys
3.37 Either order;
  a. Volcanoes
  b. earthquakes
3.38 folded
3.39 Mediterranean
3.40 Any order:
  a. Africa
  b. India
  c. Australia
  d. Antarctica
  e. South America
3.41 extends from Iceland south to the tip of Africa, then eastward into the Indian Ocean. Another ridge lies in the eastern Pacific.
3.42 plastically flowing rock
3.43 Any order:
  a. volcanoes and earthquakes that circle the Pacific
  b. mountain areas where plates descend
  c. mid-ocean ridges
  d. ocean trenches

SECTION ONE

1.1 true
1.2 false
1.3 b
1.4 b
1.5 Neptunists
1.6 Plutonists
1.7 France
1.8 conclusion
1.9 Basalt was traced to the volcanic vent from which it erupted.
1,10 a. time
b. materials
1.11 Geology is an observational science because a geologist cannot do his work in a laboratory. He must depend on clues in the rocks.
1.12 A sedimentary bed could be recognized by fossils it contains.
1.13 teacher check
1.14 teacher check
1.15 teacher check
1.16 false
1.17 false
1.18 true
1,19 true
1.20 false
1.21 b
1.22 d
1.23 c
1.24 b
1.25 c
1.26 high
1.27 surface or crust
1.28 fossils
1,29 granite
1.30 clay
1.31 lithification
1.32 clastic (detrital)
1.33 formation
1,34 member
1.35 group
1.36 Organisms are buried in sedimentary rock and fossilized. Sedimentary rocks contain fossils.
1.37 Limestone is deposited in oceans. Something happened to raise a continental shelf to the height of a mountaintop.
1.38 A break in sedimentation represents a time during which sediment was not deposited.
1.39 Sedimentary rocks form from detritus (sediment) that has been weathered from first-generation granites.
1.40 a. water chemically weathers the minerals
b. the stream transports the sediment
c. the stream winnows (separates) the different kinds of sediment
1.41 Detritus is particles of pre-existing rocks; limestone comes from organisms.
1.42 a. reducing the volume, increasing the density – squeezing the sediment grains together.
b. the growing together of grains that are in contact.
c. the bonding of grains by salts (cement) precipitated from the sea water.
1.43 a. silt
b. algal structures, coral frags, seashells, sponge spines
c. gravel
d. limestone
e. mudstone
f. conglomerate
g. sandstone
h. limestone
1.44 a. sand
b. mud
c. gravel
1.45 a. group
b. formation
c. member
d. bed
1.46 a. It must have shell, bones, etc.
b. It must be buried quickly

1.47 a. oxygen
b. bacteria
c. scavengers
1.48 a. petrifaction-replacement
b. total removal, leaving a mold.
c. filling of the mold producing a cast.
d. removal of liquids and solids, leaving carbon-distillation
e. freezing in regions of constant cold. Also mummification in peat bogs and oil seeps
1.49 Mud does not permit the flow-through of water that carries oxygen.
1.50 a. Good: sediment for rapid burial
b. Bad: scavengers-no burial
c. Good: low oxygen content
1.51 teacher check
1.52 teacher check
1.53 a. the search for oil
b. the search for minerals
c. the planning of bridges, tunnels, and dams
1.54 true
1.55 false
1.56 false
1.57 true
1.58 false
1.59 true
1.60 true
1.61 true
1.62 true
1.63 false
1.64 c
1.65 a
1.66 c
1.67 c
1.68 c
1.69 b
1.70 a. calcite
b. silica
1.71 poor
1.72 mold
1.73 turned to rock
1.74 trace
1.75 paleontology
1.76 orogeny

SECTION TWO

2.1 false
2.2 true
2.3 true
2.4 false
2.5 false
2.6 true
2.7 false
2.8 d
2.9 c
2.10 relative
2.11 correlating
2.12 Superposition
2.13 a. turbidites
b. thrust faults
2.14 dike
2.15 unconformity
2.16 All members of a graded bed formed by a turbidity current are deposited simultaneously.
2.17 In water, fine sediment settles more slowly than coarse sediment.
2.18 teacher check
2.19 a. campsite in three to twelve meters of water off La Jolla, California.
b. the seacoast town of Limani Chersoniso, built by the Romans inland.
c. Venice, Italy, sinking
d. Long Beach, California
e. temple of Jupiter Serapis near Pozzuoli, Italy, which shows signs of rise and fall, or a beach on the Baltic Sea, which is 500 meters above sea level
2.20 dike
2.21 M
2.22 orogeny
2.23 unconformity

2.24 b
2.25 d
2.26 d
2.27 c
2.28 false
2.29 true
2.30 true
2.31 true
2.32 false
2.33 false
2.34 true
2.35 d
2.36 b
2.37 d
2.38 d
2.39 b
2.40 c
2.41 c
2.42 a. temperature
b. food supply
2.43 Any order:
a. light
b. temperature
c. precipitation
2.44 varve
2.45 organic
2.46 Either order:
a. Sequoia
b. bristlecone pine
2.47 Genesis
2.48 relative
2.49 absolute
2.50 absolute
2.51 relative
2.52 relative
2.53 absolute
2.54 absolute
2.55 relative
2.56 absolute
2.57 5
2.58 2
2.59 5
2.60 3
2.61 Either order:
a. k
b. y

SECTION ONE

1.1 false
1.2 true
1.3 true
1.4 true
1.5 false
1.6 false
1.7 true
1.8 true
1.9 false
1.10 true
1.11 true
1.12 false
1.13 true
1.14 false
1.15 true
1.16 d
1.17 b
1.18 d
1.19 d
1.20 pathogenic
1.21 plants
1.22 animals
1.23 toxins
1.24 protoplasm
1.25 fungi
1.26 budding
1.27 H.T. Ricketts
1.28 Any order:
  a. spirillum
  b. bacillus
  c. coccus
1.29 a. overcrowded areas
  b. insufficient bathing facilities
  c. abundant insects
  d. unsanitary drinking water
  e. unsanitary toilet facilities
1.30 a. influenza
  b. measles
  c. polio
  d. mumps
  e. infectious jaundice
1.31 false
1.32 true
1.33 true
1.34 false
1.35 false
1.36 pus pockets
1.37 secondary infection
1.38 cleanliness
1.39 disinfectants or antiseptics
1.40 incubation
1.41 rash or cold
1.42 multiplying or reproducing
1.43 white blood cells
1.44 Any order:
  a. fever
  b. weakness
  c. digestive disturbances
1.45 Mention:
  airborne, food and water contact, excrement, animal and insect bite.
1.46 teacher check

SECTION TWO

2.1 false
2.2 true
2.3 false
2.4 typhoid
2.5 quarantine
2.6 carriers
2.7 Harm is caused by the toxin that the bacteria produce in the process of growing.

2.8

Disease	Incubation Period	Symptoms	Spread by
Bacillary Dysentery	1–7 days	Diarrhea, cramps, fever, bowel excretions show mucous, pus, blood	Contaminated food and water, carriers
Cholera	5–6 days	Severe constantly flowing diarrhea, collapse, vomiting, cramps	Polluted drinking water, warm moist climates
Food Poisoning Staphylococcus	6 hours	Intestinal disturbances, nausea, cramps, vomiting, headache, sweating	Staph bacteria in food (cream and milk foods)
Salmonella	within 6 hours	nausea, cramps, diarrhea, vomiting, headaches, sweating	Salmonella toxin in food
Botulism	6 hours	weakness, paralysis, impairs swallowing, talking and vision	Chlostridium botulinus toxin in home–canned foods
Typhoid	3–21 days	Headache, pain in body, listlessness, loss of appetite, bowel disturbances, gas and bloating in bowels, hemorrhage	Excretions, contaminated food, water, clothing, milk–carriers

2.9 false
2.10 true
2.11 false
2.12 true
2.13 twenty
2.14 whooping cough
2.15 lungs
2.16 scarlet fever
2.17 to combat the spread of T.B.

2.18

Disease	Incubation Period	Symptoms	Spread by
Bronchial Pneumonia	not given –hospitalization usually required	cold, chills, pain one side of body or chest, coughing, nausea, diarrhea	not given
Lobar Pneumonia	not given	cold symptoms, chills pain in side or chest, coughing and spitting, fever, weakness, headache and sometimes nausea, vomiting and diarrhea	not given
Tuberculosis	not given	severe persistent cough, spitting, loss of weight and appetite, night sweats, low–grade fever	contact with carriers milk or milk products
Scarlet Fever	2–4 days	chills, sore throat, nausea, vomiting, rapid pulse, high fever, headache and pinsize red spots on neck and chest: skin peels and scabs; "strawberry tongue"	contact with someone who has the disease
Whooping Cough	2–14 days	explosive type cough, common cold symptoms first, thick sticky mucous in mouth, loss of weight, dehydration	Saliva thrown by the cough of someone who has the disease

2.19 false
2.20 true
2.21 Because tetanus bacteria are found around livestock.

2.22

Disease	Incubation Period	Symptoms	Spread by
Cerebrospinal Meningitis	not given	inflammation of membranes of brain and spinal cord, cold symptoms, pinpoint rash, vomiting, fainting, stiffness in neck	germ–laden droplets from coughing and sneezing
Tetanus	7 days	pain in wound, muscles spasm, chills, fever, headache, stiffness in jaw and neck, attacks nerve tissue	bacteria in soil –invades through wound

3.1 c
3.2 b
3.3 a
3.4 b
3.5 have not
3.6 a. liver
Either order:
b. skin
c. eyes
3.7 a. soreness and dryness in the nose or throat
b. congested nasal passage-sneezing and runny nose
c. watery eyes
d. mucous nasal discharge-cough
e. fever and headache
f. lack of appetite- often pain in back and limbs
3.8 a. not known
b. fever, headache, soreness of throat, watery eyes
c. viruses
3.9 a. 1-3 days
b. chills, fever, headache
c. direct contact
3.10 a. 8-10 weeks
b. loss of appetite, headache, fatigue
c. virus
3.11 a. 12-19 days
b. fever, runny nose, watery eyes
c. direct contact
3.12 a. 14-23 days
b. rash swelling of lymph glands
c. direct contact
3.13 a. 14 days
b. blisters
c. contact with infected material in blister
3.14 a. 14-21 days
b. swelling of sides of face
c. carrier
3.15 a
3.16 h
3.17 g
3.18 f
3.19 d
3.20 b
3.21 e
3.22 c

Disease	Incubation Period	Symptoms	Spread by
Common Cold	not given	soreness, dryness of throat; nasal congestion, sneezing, nasal discharge; eyes water, voice raspy, difficult breathing; nasal discharge becomes mucousy; fever, headache; lack of appetite; pains in back and limbs	direct contact, infected droplets of nose or cough
Influenza	1–3 days	cold symptoms, sudden chills, fever headache, backache; fever 1–5 days	contact; droplets from coughing, sneezing, talking
Viral Hepatitis	8–10 weeks	loss of appetite; fatigue, headache, pain in upper–right side of abdomen caused by swollen liver, dull apathetic, yellowish color of skin.	contaminated food or fluid, blood transfusion
Measles	12–19 days	cold symptoms, eruption on lining of mouth, purplish–red rash on face, mouth, skin; then on rest of body; high fever, inflammation of eyes	direct contact by secretions of nose and throat
German Measles	14–23 days	cold symptoms, red rash; lymph glands on back of neck swell, become tender	direct contact
Chicken Pox	2 weeks	cold symptoms, watery blisters, fever, then blisters scab; scabs last 2–4 days each	infectious material in blisters, can be inhaled
Mumps	14–21 days	fever, chills, headache, loss of appetite, dry mouth, pain from chewing or swallowing, glands in front of ears swell.	saliva
Infantile Paralysis	usually 7–14 days but can be 5–35 days	cold or flu symptoms, temporary paralysis of arms or legs; permanent paralysis or death; if severe, affects spinal cord	nose/throat excretion, bowel movements, sneezing, coughing; contaminated food, water, sewage
Smallpox	8–12 days	violent headache, chills, pains, high fever; small reddened pimples form scabs; face swells; rash is painful and eyelids swell	secretions of nose, throat blisters and scabs; body excretions
Rabies	10 days–12 months; usually 20–90 days	first irritability, apprehension, tingling pain in wound; second, hoarseness, feeling choking; third, spasms in swallowing and breathing muscles; pain in swallowing; fourth, thick foaming saliva; fifth, convulsions and death.	saliva of infected person or animal
Yellow Fever	3–6 days	sudden symptoms; flushed and swollen face, dull eyes, bright red lips and tongue, high fever, pain in head and back, exhaustion; then temperature drops, skin gets cold and jaundice appears	saliva of infected person or animal

Contagion Period	Care and Treatments of Patient	Vaccination	Additional Comments
perhaps 2 weeks	bed rest, plenty of liquids, light diet, keep warm, gargle, possible aspirin; observe cleanliness procedures, proper disposal of handkerchiefs	none	not known what causes colds; no virus isolated; avoid secondary infections
not given not known	bed rest, warm, bathed, take much fluid; isolate; avoid exposure to source of secondary infection complications	few viruses isolated	secondary infections common; severe cases may need hospitalization
not given	bed rest, high carbohydrate, high protein diet, extra vitamins, 8–10 weeks for recovery	none	infects liver; affects ages 25–40 most often
as long as nose throat secretions are present	thorough washing of hands, bedding, dishes; materials must be disinfected; isolate; keep strong light away	yes	antibiotics given for secondary infections; illness provides lifetime immunity
same	bed rest; not as severe or long lasting as measles	yes	dangerous for pregnant women, danger to fetus –antibiotics for secondary infections
not known	isolate as long as scabs present; disinfect bedding; air room; apply antibiotic ointments to blisters	yes	virus causes shingles in adults; lifetime immunity
up to 2 weeks	keep quiet, mouth cleansed, warm compresses on swellings; guard against secondary infections; relapse complications	yes	gamma globulin; antibiotics given; children 5–15 years usually
not given	not given; complete bed rest; in severe cases hospitalization; consult physician	yes	most feared infectious disease today
up to 4 weeks –until scabs are gone	isolation; quarantine; disinfect clothing, bedding and all materials used; guard against secondary infections	yes, repeat 5–7 years	lifetime immunity from disease
from time symptoms appear until death	Pasteur treatment if rabid animal cannot be found and examined; Pasteur treatment if animal is rabid	for dogs, yes	controlled in England, Canada, and United States
when there are infected mosquitoes	not given	yes, for travelers	mosquito control controls the disease

SECTION FOUR

4.1 false

4.2 true

4.3 false

4.4 false

4.5 true

4.6 death

4.7 rickettsial

4.8 tick

4.9 dogs

4.10 typhus

4.11 lice or rat fleas

4.12 epidemic typhus

4.13 ringworm

4.14 athlete's foot

4.15 body ringworm

4.16 protozoan

4.17 intestinal tract/large intestine

4.18 excretions

4.19 Anopheles mosquito

4.20 tropical or semitropical

4.21 carrier

4.22 true

4.23 false

4.24 true

4.25 true

4.26 false

4.27 true

4.28 false

SECTION ONE

1.1 disease
1.2 immune or susceptible
1.3 skin
1.4 epithelial
1.5 endothelial
1.6 fibroblasts
1.7 macrophages
1.8 leukocytes or white blood cells
1.9 interferon
1.10 fever
1.11 antibodies
1.12 antigen
1.13 immunity
1.14 their mothers
1.15 a
1.16 d
1.17 a. destroying them
b. by causing them to clump so white blood cells can destroy them
1.18 Any order:
a. measles
b. tetanus
c. whooping cough
d. diphtheria
e. polio
f. mumps
1.19 teacher check
1.20 teacher check
1.21 Answer should include:
taken internally, a chemotherapeutic agent to destroy micro organisms but not cells
1.22 no
1.23 Example:
Specific antibiotics treat specific diseases;
doses too large or too small are ineffective and may be harmful.
1.24 yes
1.25 Brand name
Achromycin
Terramycin
Penicillin-G
Keflin
Use
broad-spectrum
same
same
same
1.26 All
1.27 all differ; specific allergies to drugs
1.28 Name (examples)
analgesics
vitamins
antihistamine
cough remedies
decongestants
Main ingredients
aspirin
Use
pain
deficiencies
allergies
coughs
nasal congestion
1.29 teacher check

SECTION TWO

2.1 false
2.2 false
2.3 true
2.4 true
2.5 true

2.6 overemphasis
2.7 social
2.8 microorganisms
2.9 political
2.10 Hint:
Mention health, sanitation, toilet facilities, cleansers, municipal facilities.
2.11 van Leeuwenhoek
2.12 Sabin and Salk
2.13 Edward Jenner
2.14 Harvey Cushing
2.15 Alexander Fleming
2.16 Emil Behring
2.17 Louis Pasteur
2.18 Hans Bergen
2.19 Robert Koch
2.20 Paul Ehrlich
2.21 Karl Landsteiner
2.22 Ilya Mechnikov
2.23 E. Ruska
2.24 true
2.25 false
2.26 true
2.27 true
2.28 true
2.29 true
2.30 true
2.29 true
2.31 true
2.32 Genetic characteristics are acquired from the parents by the offspring and determine our physical traits.
2.33 some illness is due to a person's emotional state and outlook on life
2.34 Nutrition is a term used to describe the study of the raw materials needed for our body to function properly.
2.35 white cells, antibodies, skin, mucous membrane, cilia
2.36 reduces disease-causing organisms

SECTION THREE

3.1 b
3.2 c
3.3 a
3.4 g
3.5 i
3.6 f
3.7 e
3.8 h
3.9 d
3.10 volunteer
3.11 efficiency
3.12 a. specialize
b. specialists
3.13 machine
3.14 based on one's own wishes, notions, or will; not going by rule or law
3.15 a. dosage
b. poisoning
3.16 Either order:
a. National Institute of Health
b. Food and Drug Administration
3.17 Any order:
a. National Cancer Institute
b. National Heart Institute
c. National Institute of Mental Health
3.18 information
3.19 Food and Drug Administration
3.20 appropriate or pertinent or accurate

SECTION ONE

1.1 false
1.2 true
1.3 false
1.4 a
1.5 b
1.6 c
1.7 For six of the known planets, Bode's Law is a very close approximation in A. U. from the actual mean distance since confirmed.
1.8 An astronomical unit is the distance of the earth from the sun, or 93,000,000 miles.
1.9 count the number of zeros in a number and write it as a superscript to the base number 10.
1.10 teacher check
1.11 false
1.12 true
1.13 false
1.14 c
1.15 d
1.16 a
1.17 A linear increase in brightness (1 mag. diff.) according to the eye is precisely measured as a geometric increase in brightness (2.51 times brighter).
1.18 sixth magnitude
1.19 The parallax technique is used to indirectly measure the distance to far away objects. The distance to the star is relative to its parallax angle, which is one-half the apparent change in its angular position during a six-month period.
1.20 a. Carina
b. Camelopardalis
c. Cassiopeia
1.21 because one is dealing in relative distances
1.22 If the difference in apparent magnitude (2.3) is rounded to 2, the answer is 6.3. There are two ways to solve the problem with more precision.
a. Plot the data on top of page 7 and then plot the magnitude difference (-1.5-(+0.8) or 2.3 on the graph. The brightness ratio will equal about 8.3.
b. The second way is to use the method shown on page 7.
$2.51^{2.3} = (2.51)^{0.1} \times (2.51)^{0.1} \times (2.51)^{0.1} \times (2.51)^{2.0}$
$= 1.1 \times 1.1 \times 1.1 \times 6.3 = 8.3$
1.23 Draco is farther away by about 1.81 LY.
Solution:
Alpha Draco: $pc = \frac{1}{parallax} = \frac{1}{1\ 0.18''}$
$= 5.55$ pc
$5.55\ pc \times 3.26\ \frac{LY}{pc} =$ <u>18.11 LY</u>

Altair: $pc = \frac{1}{parallax} = \frac{1}{0.20''}$
$= 5.00$ pc
$5.00\ pc \times 3.26\ \frac{LY}{pc} =$ <u>16.30 LY</u>

Distance difference
= 18.11 LY - 16.30 LY
= <u>1.81 LY</u>

1.24 6×10^{13} miles

SECTION TWO

2.1 aperture

2.2 one-third inch

2.3 Area of 2″ lens $= \pi\left(\frac{d}{2}\right)^2$

$$= \pi\left(\frac{2}{2}\right)^2$$

$$= \frac{\pi}{1}$$

Ratio of areas $= \frac{\pi}{36} \div \frac{\pi}{1}$

$$= \frac{\pi}{36} \left(\frac{1}{\pi}\right)$$

$$= 1:36$$

A 2-inch telescope is 36 times as powerful as the unaided eye.

2.4 d

2.5 a

2.6 f

2.7 e

2.8 h

2.9 i

2.10 chromatic aberration

2.11 flint

2.12 30X

2.13 see figure 5

2.14 see figure 6

2.15 spectrograph

2.16 Either order:
- a. strength of hydrogen line (chemical composition)
- b. surface temperature

2.17 O

2.18 Q

2.19 Arecibo

2.20 relatively small celestial object which shines brighter than a hundred normal galaxies

SECTION THREE

3.1 c

3.2 d

3.3 g

3.4 a

3.5 b

3.6 e

3.7 Either order:
- a. force of gravity
- b. centrifugal force

3.8 second

3.9 Either order:
- a. the velocity of the earth's rotation about its axis
- b. its best range of latitudes on the earth's surface for satellite observation

3.10 Any order:
- a. communications satellite
- b. weather satellite
- c. earth-resources satellite
- d. military satellite

3.11 to map the existence of certain mineral deposits and the extent of other natural resources

3.12 Evidence that a great deal of water once flowed on Mars. Mars is considered to be in the sun's habitable zone. Some forms of earth life can survive Martian environmental conditions.

3.13 Some form of activity was present in soil samples, but scientists cannot explain whether it is because of living organisms or some very unusual chemical characteristics.

3.14 that the natural laws that govern earth's physical and chemical events apply universally

SECTION ONE

1.1 i
1.2 g
1.3 h
1.4 f
1.5 a
1.6 j
1.7 e
1.8 d
1.9 b
1.10 c
1.11 He ordered the production of Gulf Stream charts which enabled faster ocean crossing for the mails between England and the colonies.
1.12 a. Its 3-year voyage brought back so much data as to interest greatly the scientific community in oceanographic research.
b. covered over 140 million square miles of ocean in 3 years
c. Its chemist, Buchanan, was credited with founding chemical oceanography.
d. Its scientific reports provide 50 volumes of data on the seas.
1.13 Example:
Haber sought gold from seawater in 1925, which was unsuccessful. However, his data on salt nutrient distribution in relation to plankton proved very beneficial.
1.14 Radio Direction Finder
1.15 subbottom profiles
1.16 4,650 feet
1.17 The current drag is submerged beneath the surface and measures currents independently of the surface float which is affected by surface winds and currents.
1.18 d
1.19 a
1.20 f
1.21 e
1.22 b
1.23 c
1.24 mid-Atlantic ridge
1.25 a. Cousteau
b. Conshelf Two
1.26 Any order:
a. uranium
b. silver
c. gold
1.27 e
1.28 g
1.29 a
1.30 f
1.31 c
1.32 b

SECTION TWO

2.1 Any order:
a. continental drift
b. sea-floor spreading
c. global tectonics
2.2 A transmitter sends a soundwave that is reflected to a receiver. From the time required for the return, the depth can be calculated; and a profile of the ocean floor can be derived.
2.3 continuous seismic profiling
2.4 seismic refraction
2.5 coring
2.6 a. hundred cores
b. 1,000
2.7 ecogram
2.8 c
2.9 f
2.10 e

2.11 b
2.12 d
2.13 Europe
2.14 pressures
2.15 a. pressure
b. 14,000
2.16 false
2.17 true
2.18 false
2.19 true
2.20 Sediments are classified according to their size, chemical composition and place of deposit.
2.21 Sea-floor sediments are increasingly thick away from the mid-ocean ridges because they have had more time to accumulate sediment.
2.22 Because they are near the western boundaries of their ocean systems.
2.23 tuna gear was observed to be moving eastward below westward moving surface currents
2.24 friction between thin water layers
2.25 d
2.26 c
2.27 a

SECTION THREE

3.1 false
3.2 true
3.3 true
3.4 false
3.5 true
3.6 copepod
3.7 its ease of capture
3.8 zooplankton
3.9 trawl net
3.10 e
3.11 a
3.12 b
3.13 f
3.14 d
3.15 c
3.16 d
3.17 a
3.18 c
3.19 b
3.20 a. hydrogen
b. oxygen
c. 96
Any order:
d. chlorine
e. 2
f. sodium
g. 1
h. magnesium
i. 0.1
3.21 c
3.22 e
3.23 d
3.24 a
3.25 b
3.26 tsunamis
3.27 gravity
3.28 Arctic Ocean
3.29 France
3.30 the petroleum embargo
3.31 16 per cent

SECTION ONE

1.1 Water droplets begin to collect.
1.2 in the top of the bag (over the "garden")
1.3 Either order:
  a. evaporation
  b. precipitation
1.4 true
1.5 true
1.6 true
1.7 71.7% or 72%
1.8 19%
1.9 30.1% or 30%
1.10 false
1.11 false
1.12 true
1.13 70
1.14 1
1.15 mineral
1.16 desalinated
1.17 extinction
1.18 Example:
  death to animals and plants;
  oil stained beaches
1.19 The soil began to settle to the bottom.
1.20 Some small particles remained floating in the water.
1.21 Answer depends on amounts of dirt and water used.
1.22 Shaking polluted the environment.
1.23 true
1.24 true
1.25 recycling
1.26 ultrasonic
1.27 dispersant
1.28 It provides a drain from the high to the low.
1.29 The sand was channeled and rocks were moved.
1.30 Answer may mention migration routes, scenery, breeding or feeding areas.
1.31 Either order:
  a. The reservoir was higher.
  b. The reservoir had more water.
1.32 false
1.33 true
1.34 the smallest
1.35 yes
1.36 the smallest container with the most drops of color
1.37 Answer will depend on sizes of containers; smaller container should take less time.
1.38 add more water
1.39 false
1.40 false
1.41 true
1.42 teacher check
1.43 population
1.44 five
1.45 5
1.46 famine
1.47 a. help
  b. need
1.48 a. relief
  b. grain
  c. India
1.49 meats, soybeans, peanut butter
1.50 yes
1.51 no
1.52 yes
1.53 true
1.54 false
1.55 true
1.56 50 million tons
1.57 100 million tons
1.58 250 million tons
1.59 more, because the amount of tons per year is higher and shows a greater slope on the graph
1.60 true
1.61 false
1.62 Examples:
  paper bags, plastic cups, wooden spoons

1.63 Examples:
cans, plastic containers, bags
1.64 Examples:
paper, plastic, metal
1.65 refined
1.66 recycling
1.67 about 6 billion people
1.68 5.9%
1.69 Asia
1.70 U.S.S.R.
1.71 false
1.72 false
1.73 false
1.74 true
1.75 false
1.76 true
1.77 true
1.78 false

SECTION TWO

2.1 metal; food products
2.2 plastic
2.3 wear gloves; have water present; avoid splashing; and avoid inhaling vapors
2.4 water dilutes the acid (makes the acid less effective)
2.5 true
2.6 false
2.7 false
2.8 Examples:
electricity, human muscle, battery, wind, solar
2.9 Examples:
electricity - lights
human muscle - work accomplished
battery - movement; sound
wind - movement
solar - heat
2.10 Examples:
a. electricity
b. solar
c. solar
2.11 Any order:
a. depleted natural resources
b. destroyed landscape through poor mining methods
c. failed to restore the balance of nature
2.12 Any order:
a. oil
b. natural gas
c. coal
2.13 oil-based
2.14 a. 75
b. 80
c. 95
2.15 sulfuric acid
2.16 ash
2.17 Either order:
a. emphysema
b. black lung
2.18 fossil
2.19 fossil
2.20 energy
2.21 fission
2.22 Roentgens or R's
2.23 500
2.24 uranium 235
2.25 false
2.26 false
2.27 false
2.28 false
2.29 true
2.30 true
2.31 a
2.32 d
2.33 c
2.34 fusion
2.35 centuries
2.36 technical complexity
2.37 uranium
2.38 liquid
2.39 true
2.40 true
2.41 true

2.42 false
2.43 Hint:
Measure the length and the width, and multiply.
2.44 Hint:
Look at the publisher's comment, usually inside the front page.
2.45 Hint:
Multiply the square-foot coverage of one paper times the circulation number.
2.46 Any order:
a. car pools
b. rapid transit systems
c. technological advances
2.47 waste
2.48 United States Army Corps of Engineers
2.49 sandstone 300 feet thick
2.50 the paper honeycomb
2.51 true
2.52 false
2.53 true
2.54 false
2.55 true
2.56 Los Angeles
2.57 finances
2.58 hydrofoil
2.59 urban
2.60 Industrial
2.61 employment
2.62 suburbs
2.63 taxes
2.64 melting pot

SECTION THREE

3.1 heat shield
3.2 solar
3.3 Any order:
a. air
b. waste
c. water
3.4 earth
3.5 true
3.6 true
3.7 false
3.8 true
3.9 false
3.10 false
3.11 true
3.12 b
3.13 c
3.14 a
3.15 d
3.16 a
3.17 c
3.18 Any order:
a. silica-fiber insulation
b. laser-beam advances
c. sapphire fiber of high tensile strength
3.19 the ocean depths
3.20 inaccessibility of ocean depths
3.21 food (i.e., protein)
3.22 to discover natural resources to supply energy needs
3.23 Any order:
a. iodine
b. copper
c. vanadium
3.24 false
3.25 true
3.26 false
3.27 false
3.28 95
3.29 Either order:
a. oil
b. natural gas
3.30 concentrate

3.31 a. drugs
b. anesthetics
3.32 iodine
3.33 b
3.34 c
3.35 a
3.36 d
3.37 b
3.38 true
3.39 false
3.40 true
3.41 true
3.42 true
3.43 true
3.44 true
3.45 true
3.46 Either order:
a. energy
b. minerals
3.47 243 million
3.48 10
3.49 Either order:
a. nuclear
b. solar
3.50 nuclear
3.51 a. way
b. truth
c. life
3.52 Isaiah

SECTION ONE

1.1 a. meter
b. milliliter or liter
c. meter
d. centimeter
e. liter
f. centimeter or millimeter
g. meter
h. milliliter
i. metric ton
j. kilometers
k. milliliters
l. meter
m. centimeter
n. meter
o. liter

1.2 a. 4.54 kilograms
b. 2.7 meters
c. 27 grams
d. 3.8 liters
e. .454 kg or 454 gram loaf
f. 57 liters
g. 3.2 kilometers
h. 76.2 centimeters
i. 1.8 meters
j. 10.16 centimeters
k. 4.5 meters
l. 96 kilometers
m. 1.9 liter
n. 66.04 centimeters
o. 3 meters
p. 12.8 kilometers

1.3 a. more
b. more
c. less
d. more
e. less
f. more
g. more
h. less
i. more
j. less

1.4 Examples:
a. Astronauts had to strap themselves in bed.
b. Food floated off the plates.
c. Items came out of drawers when they were opened.
d. Using tools was difficult because their entire body moved.
e. Walking on the floor was difficult.

SECTION TWO

2.1 d
2.2 e
2.3 a
2.4 b
2.5 g
2.6 c
2.7 a. They were bankrupt and too many men were dying.
b. May, 1904
c. Walter Reed
2.8 Typhus is spread by fleas, especially by fleas found on rats.
2.9 Swimmer's itch is a condition related to schistosomiasis that is common in some lakes of the United States.
2.10 teacher check
2.11 Any order:
a. skin
b. secretions
c. white blood cells
d. antibodies
2.12 1940's
2.13 1966

2.14 World Health Organization
2.15 the disease leaves raised scars on on the face and body
2.16 Either order:
a. yellow fever
b. malaria
2.17 Either order:
a. rats
b. fleas
2.18 Any order:
a. cholera
b. amoebic dysentery
c. bacillary dysentery
2.19 the county health department
2.20 They keep down the number of infectious organisms that cause disease.
2.21 Either order:
a. you do not get the disease
b. you cannot spread the disease
2.22 A mosquito bites a person with malaria and then carries the protozoan to a new individual
2.23 Resistant forms of mosquitoes, fleas, and rats have been developed.
2.24 a change in the diet or different minerals in the water
2.25 The water used in making the ice is probably not pure.
2.26 a. by immunization, mosquito net, long clothing, or insect repellent
b. immunization, boil or treat all water, eat only freshly cooked food
c. immunization, avoid unsanitary conditions where rats live
d. immunization if the disease ever returns
e. antimalarial medication before, during, after visit; mosquito net
f. not swimming or washing in unsanitary waters that might carry the parasite
g. drink only purified water; eat only cooked or peeled foods
h. probably every traveler will have a few bouts with diarrhea, if it persists a doctor should be consulted since it might not be diarrhea but dysentery or cholera
2.27 teacher check
2.28 a. Keep foods hot or cold as required.
b. Remove ticks daily.
c. Have tetanus injection before going on trip.
d. Keep dry; watch for storms; and carry warm clothing, high calorie food, and rain gear.
e. Avoid suspicious animals and have injections if necessary.
f. Boil or chemically purify all drinking water.
2.29 teacher check
2.30 c
2.31 a
2.32 b
2.33 e
2.34 f
2.35 Either order:
a. boil it
b. treat it with purifying tablets
2.36 tick
2.37 food poisoning
2.38 puncture wounds
2.39 Any order:
a. sufficient rest
b. exercise
c. proper food
d. medical care
e. immunizations
2.40 Carry plenty of warm clothing, high-energy food, and rain gear
2.41 the family
2.42 Capture and observe the animal for fourteen days. Have rabies injections if the animal cannot be found or if it has rabies.
2.43 What? Know ye not that your body is the temple of the Holy Ghost which is in you, which ye have of God, and ye are not your own? For ye are bought with a price: therefore glorify God in your body, and in your spirit, which are God's.

SECTION THREE

3.1 erosion of existing mountains and glaciers
3.2 Any order:
a. volcanoes, any volcano; Hawaiian Islands
b. folds, mountains
c. faults, earthquakes
d. sedimentation, behind dams, river deltas
3.3 folding of sedimentary rock
3.4 San Andreas Fault in California
3.5 A fold just lifts up; a fault breaks and some material is moved out of place.
3.6 wind, rain (water), glaciers
3.7 in layers in the ocean at mouths of rivers
3.8 teacher check
3.9 Example:
The river is constantly depositing more material and increasing the size of the delta. In the past the delta was smaller because less silt and soil was deposited.
3.10 The military needed the information for submarines.
3.11 The ocean is deep and dark.
3.12 new instruments
3.13 a high ridge with a valley down the center
3.14 It is getting wider.
3.15 Either order:
a. alternating bands of magnetism
b. Readings on each side of the ridge were mirror images of each other.
3.16 along trenches and midocean ridges
3.17 increasing by 2.5 cm a year
3.18 Any order:
a. fossils
b. lava flows
c. glaciers
d. rock layers
3.19 the continental shelf
3.20 It is being formed in the valleys of the midocean ridge.
3.21 It forces them farther apart.
3.22 The ocean floor nearest the ridge is the newest and sediments have not had more time to collect.
3.23 The floor is older as one approaches the continents so sediments have had more time to collect.
3.24 Teacher check
3.25 ellipse
3.26 galaxy
3.27 billion
3.28 one billion
3.29 six trillion
3.30 parsec
3.31 2,500
3.32 constellations
3.33 seventeenth
3.34 refracting
3.35 lenses
3.36 Any order:
a. planets
b. satellites
c. planetoids
d. comets
3.37 Either order:
a. spectrograph
b. photographic film
3.38 They receive radiation beyond the frequencies of visible light.
3.39 Either order:
a. to place radio telescopes above the earth's opaque atmosphere
b. to collect information from the moon and planets

SECTION FOUR

4.1 F, fluorine, 9, 10, 19, 9
4.2 Si, silicon, 14, 14, 28, 14
4.3 Cr, chromium, 24, 28, 52, 24
4.4 O, oxygen, 8, 8, 16, 8
4.5 Au, gold, 79, 118, 197, 79
4.6 Ne, neon, 10, 10, 20, 10
4.7 Ca, calcium, 20, 20, 40, 20
4.8 Pt, platinum, 78, 119, 197, 78

4.9 a.

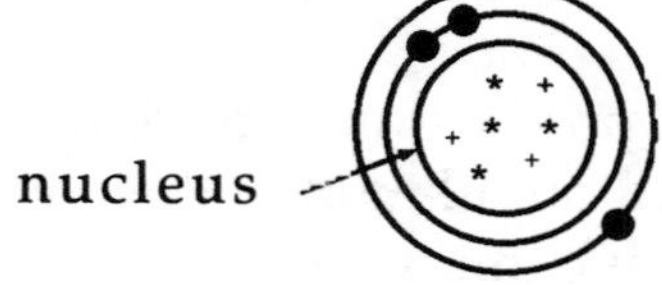

Lithium
3 electrons
3 protons
4 neutrons

Note: * = Neutron
+ = Protron
• = Electron

b.

nucleus

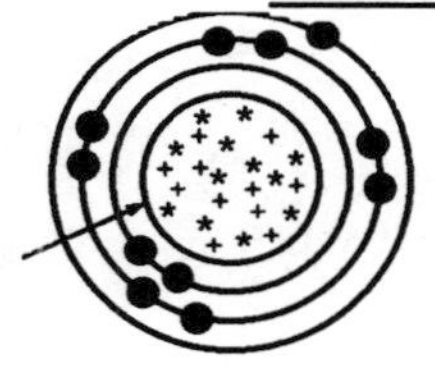

Sodium
11 electrons
11 protons
12 neutrons

c.

nucleus

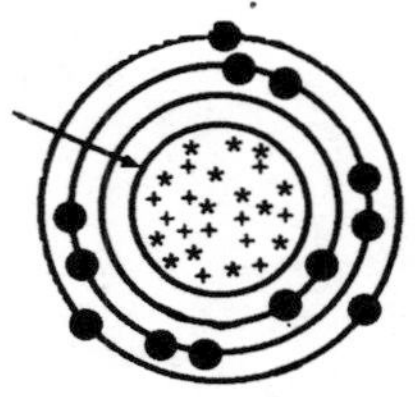

Aluminum
13 electrons
13 protons
14 neutrons

Note: * = Neutron
+ = Protron
• = Electron

d.

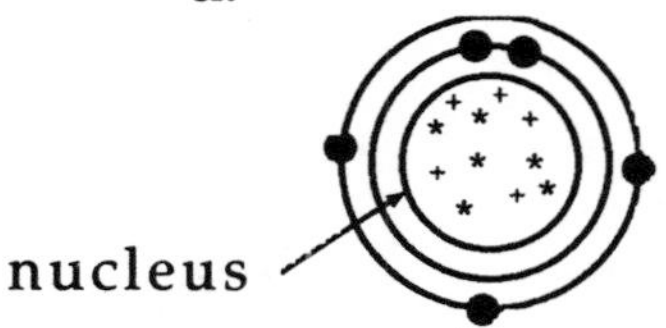

Boron
5 electrons
5 protons
6 neutrons

e.

nucleus

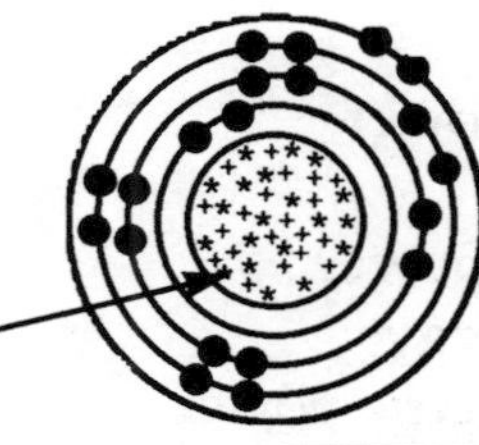

Calcium
20 electrons
20 protons
20 neutrons

Note: * = Neutron
+ = Protron
• = Electron

f.

nucleus

Fluorine
9 electrons
9 protons
10 neurons

4.10 transportation, manufacturing farming, recreation
4.11 United States
4.12 one-third
4.13 Too many environmental and safety problems exist.
4.14 Fusion creates less radioactivity.
4.15 Currently no way has been developed to control fusion
4.16 a neutral particle found in the nucleus of an atom
4.17 a positively-charged particle found in the nucleus
4.18 a tiny negatively-charged particle circling the nucleus
4.19 the number of protons in an atom
4.20 the dense central portion of an atom, composed of protons and neutrons
4.21 the mass of the protons plus the neutrons
4.22 a form of an element differing only in the number of neutrons
4.23 the splitting of an atom into lighter atoms and energy
4.24 the joining of two light atoms; produces a heavier atom and energy

4.25 teacher check
produces a heavier atom and energy
4.26 Any order:
a. atomic power
b. wind
c. thermal power
d. solar energy
e. hydroelectric power
4.27 wind is not constant or predictable
4.28 Too many electrical gadgets are in use.
4.29 It is inexpensive and nonpolluting.
4.30 Dams destroy waterways and wildlife forever.
4.31 Our government thought the area should be preserved for its beauty so all citizens could enjoy it.
4.32 Solar energy is nonpolluting, simple, and is used on the homesite.
4.33 Solar energy could conserve fossil fuels so they could be used in other areas that were colder or too cloudy.
4.34 one-fourth
4.35 cities
4.36 prime agricultural land
4.37 Any order:
a. purifying water
b. sewage treatment
c. lack of food
d. lack of medicine
4.38 Woe unto them that join house to house, that lay field to field, til there be no place, that they may be placed alone in the midst of the earth!
4.39 areas with the least food and economic potential
4.40 Each person has a smaller share of everything.
4.41 They lack fertile land, minerals, good climate, or technology.
4.42 Countries seek ways to obtain additional land or natural resources.
4.43 pollution
4.44 to increase food production to feed the expanding world population
4.45 Since humans have the same physiological needs as animals, actions which harm animals will also harm us.
4.46 a thick haze of chemicals and dust particles from industry and auto exhaust
4.47 one per cent
4.48 oxygen
4.49 a. dumping sewage
b. spill oil
c. take too many fish
d. kill whales
4.50 teacher check

Self Test 1

1.01 a
1.02 b
1.03 b
1.04 a
1.05 c
1.06 a
1.07 c
1.08 b
1.09 a
1.010 c
1.011 Any order:
- a. solid
- b. liquid
- c. gas

1.012 back and forth
1.013 Either order:
- a. back and forth
- b. turning around

1.014 unrestricted
1.015 H=1

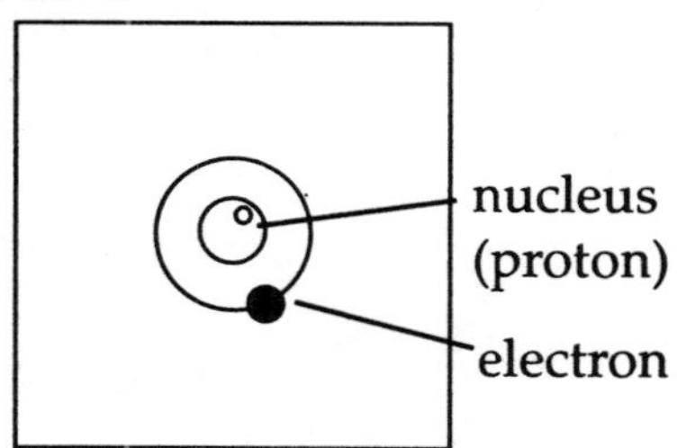

1.016 C=6

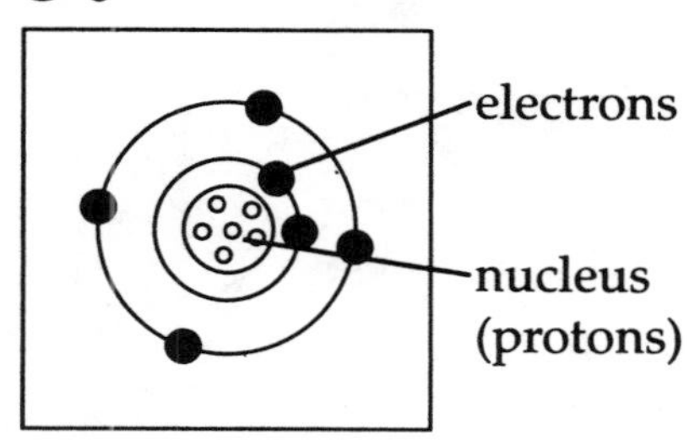

1.017 Na=11

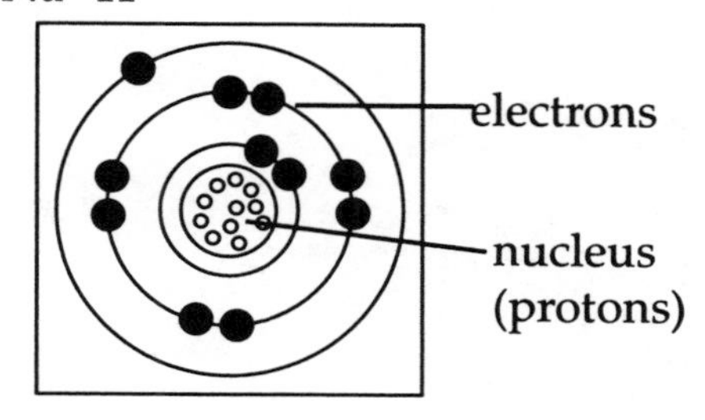

1.018 S=16

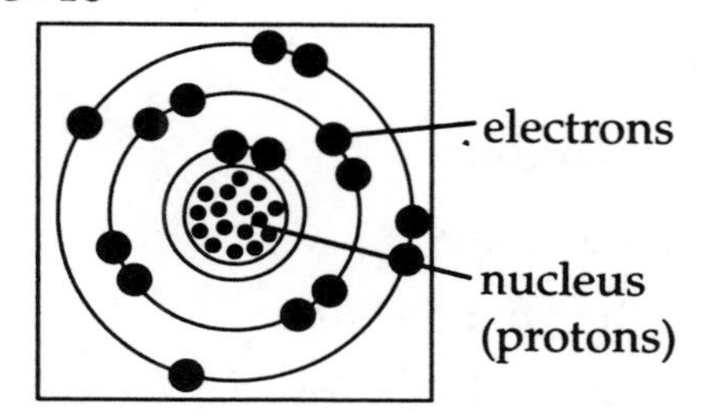

1.019 Examples; any order:
- a. A solid has mass and takes up space.
- b. A solid has definite shape and size.
- c. It is not free to move around and stay put when set down.

1.020 Examples; any order:
- a. They can move back and forth and turn around.
- b. They have no definite shape.
- c. Liquids have mass and density.

1.021
- a. Gases have unrestricted movement.
- b. Gases take shape of container.
- c. Gases have mass and low density.

Self Test 2

2.01 d

2.02 b

2.03 e

2.04 g

2.05 a

2.06 b

2.07 e

2.08 h

2.09 b

2.010 f

2.011 Any order:

a. has mass

b. has fixed shape

c. has volume (or restricted movement)

2.012 Either order:

a. Geiger counter

b. Wilson cloud chamber

2.013 atomic number

2.014 26

2.015 An accidental exposure of photographic film to radioactive materials caused the film to be exposed. Becquerel discovered through further research that the unseen rays were causing the photographic exposure.

2.016 Suggested answers: Photographic film is exposed by radioactive materials. Magnetic fields deflect the particles given off by radioactive materials. A Wilson cloud chamber is a chamber of vapor. When radiation passes through the chamber, it interacts with the vapor and leaves a vapor trail that scientists can use to study the mass and speed of the radiation going through. The Geiger counter counts radiation passing through an area of paper.

2.017 a. Beta particles are bent sharply toward the North Pole.

b. Alpha particles are slightly toward the South Pole.

c. Gamma rays are not affected.

2.018 Na=11

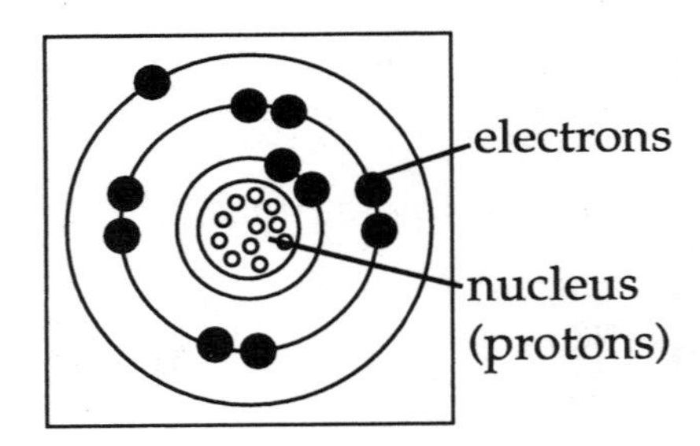

2.019 C=6

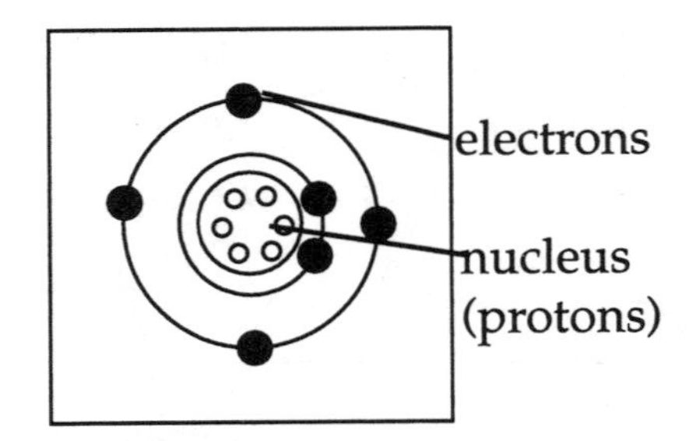

Self Test 3

3.01 h

3.02 j

3.03 e, f

3.04 g

3.05 d

3.06 b

3.07 k

3.08 e, f

3.09 a

3.010 c

3.011 a. C=6

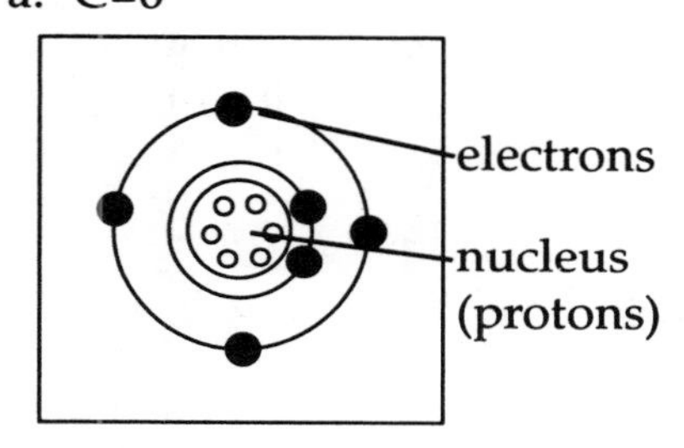

b. Ne=10

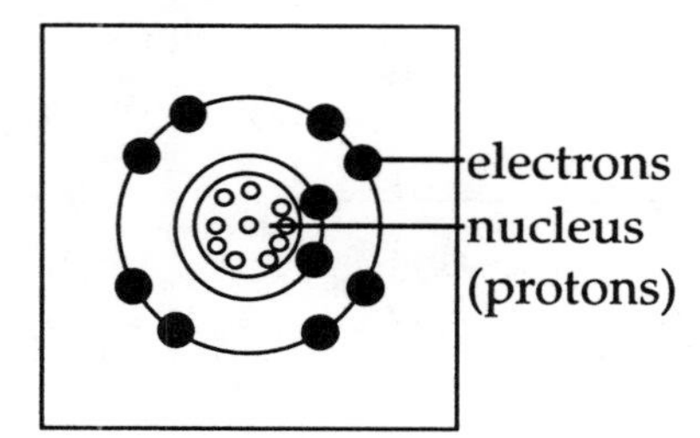

3.012 Any order:

a. proton

b. neutron

c. electron

d. positron

e. neutrino

f. gamma ray

g. meson

h. alpha

3.013 The nucleus is a series of energy levels. When neutrons and/or protons fall to a lower level, a gamma ray is emitted.

3.014 Two basic methods detect radiation – photographic film and magnetic fields. The Geiger counter and Wilson cloud chamber are two instruments used to measure the amount and intensity of radiation.

3.015 a. electron = small mass, negative charge circles atom in levels around nucleus

b. proton = 2,000 times mass of electron, positive charge, located in nucleus

c. neutron = no charge, equals mass of proton, located in nucleus

3.016 a. solid – specific shape, size and mass; molecules have restricted motion which is basically back and forth

b. liquid – free to move and flow; has mass, takes shape of container, occupies space, particles move freely but are restricted to back and forth and rotation

c. gas – no restriction on movement, fills entire container, has mass, low density

Self Test 4

4.01 d

4.02 f

4.03 h

4.04 a

4.05 j

4.06 b

4.07 i

4.08 e

4.09 g

4.010 c

4.011 back and forth and turning around

4.012 Any order:
- a. protons
- b. neutrons
- c. electrons

4.013 Any order:
- a. solid
- b. liquid
- c. gas

4.014 a. repel
- b. attract

4.015 gas

4.016 a. control rod; made of cadmium or boron steel. Used to control the rate of neutron flow; absorb free neutrons to prevent chain reaction.
- b. coolant outlet; to keep reactor cool flows through and absorbs heat for use in generating steam for electricity.
- c. shielding; made of concrete and lead. Protects environment from dangerous radiation.
- d. fuel rods (fissionable material); Example: contains fissionable material – enough to cause a nuclear reaction but less than critical mass point.
- e. moderator; placed between fuel rods to slow down but not stop the neutrons produced in decay process. Usually graphite.

4.017 H=1

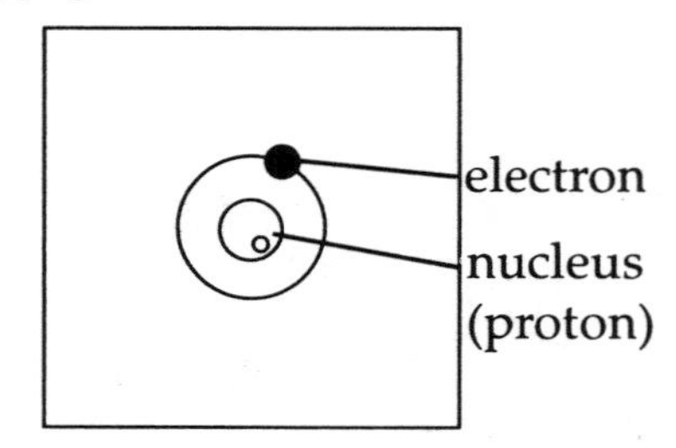

4.018 B=5

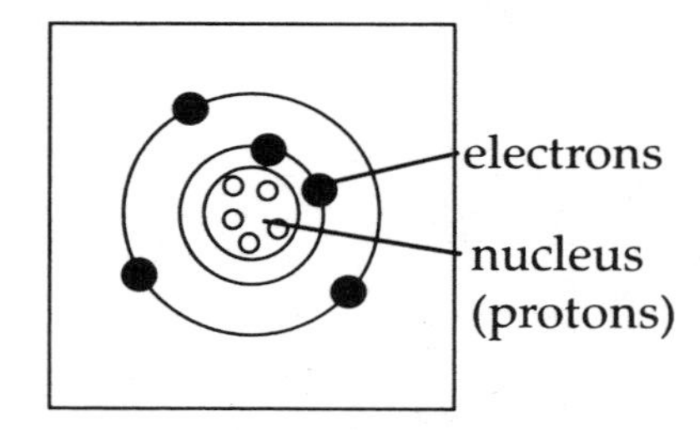

4.019 O=8

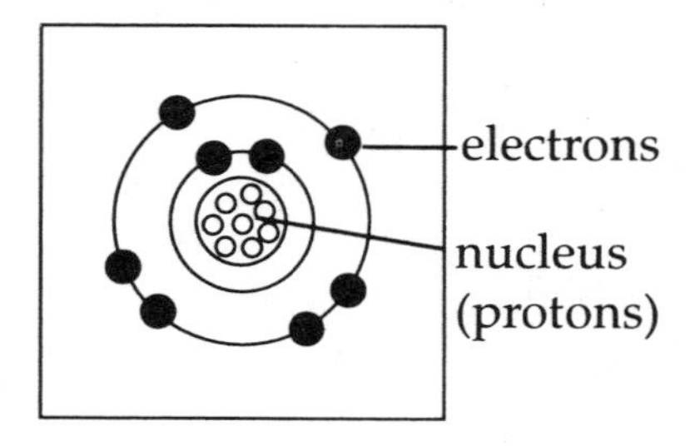

4.020 Ar= 18

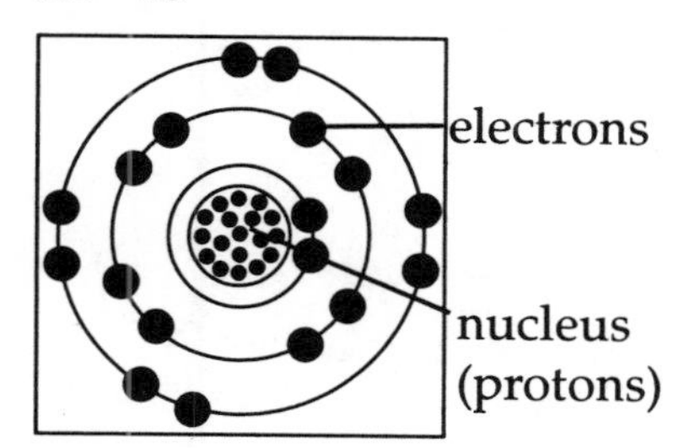

4.021 Becquerel placed radioactive material on unexposed photographic film. The material exposed the film. He continued to experiment with the effects of radioactive materials on photography film.

4.022 He identified and named some elements, headed group that built first nuclear reactor. Developed science of irradiation and bombardment.

4.023 One model has the nucleus as a series of levels in which protons and neutrons are located. The movement of a p or n down the levels emits radiation (gamma). Several particles are subatomic to a nucleus and contain positrons (positive electrons). Neutrinos (zero particles of little mass), and particles ($^{4}_{2}$He or helium nuclei).

4.024 Fission breaks down into smaller masses; fusion builds up heavier nuclei from light ones. Both release energy and lose mass: Fission is uncontrollable; fusion is uncontrollable.

Self Test 5

5.01 e

5.02 g

5.03 j

5.04 h

5.05 d

5.06 i

5.07 a

5.08 c

5.09 b

5.010 f

5.011 Either order:

a. curie

b. roentgen

5.012 Any order:

a. proton

b. neutron

c. electron (beta)

d. neutrino

e. positron

f. alpha

g. meson

5.013 Any order:

a. solid

b. liquid

c. gas

5.014 a. solid: retains shape, has mass, limited particle movement, occupies space

b. liquid: occupies space, has mass, free to flow, takes shape of container

c. gas: unrestricted movement, occupies entire container, has mass, occupies space

5.015 a. f
b. b
5.016 a. e
b. a
5.017 a. d
b. c
5.018 To treat illness (chemotherapy)
to detect illness (x-ray)
5.019 Keeps it in concentrated solid form
and buried in salt mines because it is
dangerous to people and environment.
5.020 Either order:
a. produce heat (thermal effect)
b. produce radiation
5.021 increased population and increased
consumption

Self Test 1

1.01 b

1.02 a

1.03 b

1.04 b

1.05 a

1.06 a. easily used
b. accepted by most people
c. will measure many different volumes

1.07 displacement

1.08 a. 10
b. 0.1
c. 0.062
d 1,020
e. 0.912
f. 550

1.09 1. Measure a volume of water
2. Add the coal
3. Measure the new volume
4. Subtract the first volume from the last to get volume of coal.

1.010 It does not take into account all the factors that determine volume, such as air space and gas compressibility.

Self Test 2

2.01 false

2.02 true

2.03 false

2.04 false

2.05 true

2.06 true

2.07 true

2.08 true

2.09 true

2.010 false

2.011 liter

2.012 kilogram

2.013 Closeness with which a measurement compares to an actual value.

2.014 How consistently a measurement reproduces the same value.

2.015 1000

2.016 1000

2.017 Answer should include:
1. Basic construction details
2. How to balance the rod
3. How to balance (weigh) an object by using a standard mass unit.

2.018 To insure uniformity and consistency from time to time and place to place.

Self Test 3

3.01 e

3.02 f

3.03 d

3.04 j

3.05 g

3.06 a

3.07 c

3.08 b

3.09 k

3.010 h

3.011 The buoyant force exerted by a liquid is exactly equal to the mass of the liquid displacement.

3.012 An accepted unit of measurement which is convenient to use, easily accessible, and accepted by most people. Without a standard, consistency and comparisons would be very difficult.

3.013 Mass is a measure of the "stuff" (matter) in a substance and this is a constant value any place in the universe. Weight is a measure of the gravitational attraction of one body for another and changes as location and bodies change.

3.014 Density is a measure of the "packedness" of a substance and has units of g/cm^3. Specific gravity is a ratio of densities and has no units of measurement.

3.015 Accuracy is a measurement is to the actual value. Precision is a measure of how consistently the same measurement can be reproduced

3.016 liter

3.017 one

3.018 mass

3.019 weight

3.020 a. phase
b. temperature
c. pressure

3.021 $D = \frac{8.2\ g}{3.6 cm^3} = 2.3\ g/cm^3$

3.022 $D = \frac{38.6\ g}{2.0\ cm^3} = 19.3\ g/\ cm^3$

3.023 $S.G. = \frac{16.2\ g}{8.1\ g} = 2.0$

3.024 $S.G. = \frac{96.2\ g}{12.6\ g} = 7.6$

3.025 Example:
1. Basic features of an equal-arm balance.
2. Basic ways to make a balance balance.

Self Test 1

1.01 e
1.02 k
1.03 i
1.04 g
1.05 a
1.06 j
1.07 b
1.08 h
1.09 f
1.10 d
1.11 false
1.12 false
1.13 true
1.14 false
1.15 true
1.16 true
1.17 false
1.18 false
1.19 true
1.20 false
1.21 b
1.22 b
1.23 d
1.24 a
1.25 c
1.26 c
1.27 d
1.28 b
1.29 a
1.30 d
1.31 volcanic shield
1.32 weakness
1.33 sea mounts or guyots
1.34 batholith
1.35 granite
1.36 meteorites
1.37 core
1.38 nitrogen
1.39 metamorphic
1.40 sandstone

Self Test 2

2.01 f
2.02 d
2.03 a
2.04 g
2.05 b
2.06 j
2.07 e
2.08 k
2.09 h
2.10 i
2.011 false
2.012 true
2.013 false
2.014 true
2.015 true
2.016 true
2.017 false
2.018 false
2.019 false
2.020 true
2.021 c
2.022 a
2.023 b
2.024 a
2.025 d
2.026 d
2.027 a
2.028 b
2.029 b
2.030 b
2.031 flood plain
2.032 meanders
2.033 dry or desert
2.034 erratic(s)
2.035 turbidity
2.036 shelf
2.037 river
2.038 calving

2.039 water vapor
2.040 sedimentary
2.041 Whether rock material is transported and then deposited. The deposited rock fragments can be back into rock.

Self Test 3

3.01 e
3.02 a
3.03 h
3.04 g
3.05 j
3.06 i
3.07 k
3.08 d
3.09 c
3.010 b
3.011 false
3.012 false
3.013 true
3.014 true
3.015 true
3.016 true
3.017 false
3.018 false
3.019 true
3.020 true
3.021 a
3.022 c
3.023 c
3.024 a
3.025 a
3.026 d
3.027 a
3.029 b
3.030 d
3.031 density
3.032 sedimentary
3.033 plateau
3.034 volcanoes
3.035 southern
3.036 igneous
3.037 magma
3.038 lava
3.039 oxbow
3.040 weathering
3.041 density $= \frac{\text{mass}}{\text{volume}} = \frac{3.6\text{g}}{1.2\text{ cm}^3} = 3\ \frac{\text{g}}{\text{cm}^3}$
3.042 Sp. gr. $= \frac{\text{mass in air}}{\text{mass in air} - \text{mass in water}} =$

$$\frac{160\text{g}}{160\text{g}-100\text{g}} = \frac{160\text{g}}{60\text{g}} = 2.67$$

Self Test 1

1.01 e
1.02 d
1.03 f
1.04 f
1.05 j
1.06 k
1.07 h
1.08 b
1.09 c
1.010 i
1.011 true
1.012 false
1.013 true
1.014 true
1.015 false
1.016 true
1.017 false
1.018 false
1.019 false
1.020 false
1.021 true
1.022 true
1.023 true
1.024 true
1.025 true
1.026 a
1.027 c
1.028 a
1.029 b
1.030 b
1.031 d
1.032 a
1.033 b
1.034 a
1.035 b
1.036 Because sedimentary rocks are made of fragments of other rocks.
1.037 a. sandstone
b. conglomerate
c. siltstone
d. claystone
e. limestone

Self Test 2

2.01 d
2.02 e
2.03 g
2.04 h
2.05 a
2.06 b
2.07 c
2.08 j
2.09 k
2.010 i
2.011 a
2.012 a
2.013 b
2.014 c
2.015 a
2.016 true
2.017 false
2.018 false
2.019 true
2.020 false
2.021 false
2.022 true
2.023 true
2.024 true
2.025 false
2.026 clastic
2.027 tree rings
2.028 quartz
2.029 petrifaction
2.030 lithification
2.031 Example:
The old Roman inland town of Limani Chersonisos is now on the seacoast.
2.032 teacher check
2.033 U
2.034 D
2.035 A
2.036 unconformity
2.037 Superposition
2.038 thrust

Self Test 1

1.01 something that causes disease
1.02 a living thing too small to be seen with the unaided eye
1.03 one-celled plants
1.04 one-celled animals
1.05 smallest organism; unable to reproduce outside a living cell
1.06 simple plants – parasitic, live in warm, damp places
1.07 pathogenic microorganism smaller than bacteria
1.08 fever
1.09 weakness
1.010 digestive disturbance
1.011 loss of appetite
1.012 runny nose
1.013 coughing (also irritability)
1.014 true
1.015 true
1.016 true
1.017 false
1.018 true
1.019 true
1.020 true
1.021 false
1.022 true
1.023 Fever can aid healing by destroying microorganisms or inhibiting their growth.
1.024 a
1.025 c
1.026 d

Self Test 2

2.01 f
2.02 a
2.03 g
2.04 d
2.05 c
2.06 b
2.07 i
2.08 e
2.09 h
2.010 j
2.011 k
2.012 l
2.013 e
2.014 c
2.015 h
2.016 i
2.017 d
2.018 g
2.019 f
2.020 b
2.021 a
2.022 j
2.023 pathogenic
2.024 bacteria
2.025 protozoa
2.026 viruses
2.027 fungi
2.028 budding
2.029 insects
2.030 Any order:
a. fever
b. weakness
c. digestive track disturbances
d. loss of appetite
or runny nose
coughing
irritability

Self Test 3

3.01 e
3.02 c
3.03 a
3.04 g
3.05 i
3.06 h
3.07 b
3.08 f
3.09 k
3.010 d
3.011 f
3.012 c
3.013 i
3.014 d
3.015 a
3.016 k
3.017 b
3.018 h
3.019 g
3.020 e
3.021 true
3.022 false
3.023 false
3.024 b
3.025 c
3.026 a
3.027 antibiotics
3.028 large intestine
3.029 pneumonia
3.030 tetanus
3.031 brain or spinal cord
3.032 pathogenic
3.033 secondary
3.034 Any order:
a. bacteria
b. virus
c. protozoa
d. fungi
e. rickettsia

3.035 Any order:
a. fever
b. digestive disturbances
c. weakness
d. loss of appetite
e. runny nose
or coughing
irritability
3.036 Any order:
a. air
b. food and water
c. sneeze mist
d. excrement
e. polluted water
or insect bite
excrement touch

Self Test 4

4.01 a. a single–celled plant or animal that produces disease
b. single–celled plants, existing in three shapes; some are harmful, others are beneficial
c. single–celled animals
d. plants lacking flowers, leaves, or chlorophyll that are fed from organic matter
e. pathogenic organisms smaller than bacteria, transmitted by insects

4.02 Any order:
a. fever
b. weakness
c. loss of appetite
d. rash or eruption on the body
e. sore throat – mild cold
4.03 Any order:
a. direct contact
b. through water, food, and air
c. insects

4.04 Any order:
  a. insects
  b. contaminated food
  c. contaminated water
4.05 Any order:
  a. food
  b. water
  c. insects
4.06 Any order:
  a. direct contact
  b. poor hygiene
  c. transfer from combs and brushes

4.07 a
4.08 e
4.09 f
4.010 h
4.011 c
4.012 b
4.013 d
4.014 e
4.015 c
4.016 a
4.017 g
4.018 i
4.019 h
4.020 b
4.021 f
4.022 k
4.023 d
4.024 Pasteur
4.025 liver
4.026 antibiotics
4.027 toxins
4.028 food poisoning
4.029 false
4.030 true
4.031 false
4.032 true
4.033 false

SELF TEST 1

1.01 Keeps organisms from entering if not broken. Acidic secretion of skin creates an unfavorable en vironment for microbe growth.
1.02 Secrete mucus that traps organisms. Cilia, tears, earwax, and urine keep organisms out.
1.03 in connective tissue—form scar tissue and wall off or surround foreign substances and parasites.
1.04 cells that eat, envelop and digest foreign substances.
1.05 in blood and lymph systems, ingest and digest foreign particles/wall off foreign particles.
1.06 speeds action of white cells and antibody formation—inhibits growth and multiplication of infectious organisms.
1.07 specific proteins that fight a specific microorganism. Antibodies destroy or cause them to clump together so they can be destroyed by leukocytes.
1.08 Produced by the body when attacked by infectious organisms such a measles, German measles, and hepatitis.
1.09 none
1.010 ✓
1.011 none
1.012 none
1.013 ✓
1.014 ✓
1.015 ✓
1.016 none
1.017 ✓
1.018 ✓
1.019 none
1.020 none
1.021 ✓
1.022 ✓
1.023 ✓
1.024 ✓
1.025 ✓
1.026 ✓
1.027-1.028 Any two:
1.027 penicillin, aureomycin, streptomycin
1.028 chlortetracyclene—or others from druggist
1.029 Man is not susceptible to most parasites that infect plants and animals.
1.030 Chemicals taken internally to destroy microorganisms, such as quinine, sulfanilamides, isoiazid, PAS, nitrofurans.
1.031 Substances produced by bacteria, fungi, or plants to suppress growth and multiplication of bacteria. Usually developed for one particular bacteria.
1.032 fights—inhabits—fungus growth
1.033 Examples; Any order:
a. Lysol
b. Phisoderm
c. boric acid
d. merthiolate
e. iodine

SELF TEST 2

2.01 h
2.02 g
2.03 d
2.04 a
2.05 f
2.06 b
2.07 e
2.08 a. defense systems
b. attitudes
c. cleanliness
d. immunizations
e. nutrition
f. genetic characteristics
g. spiritual health
2.09 true
2.010 false
2.011 false
2.012 true
2.013 true
2.014 true
2.015 false
2.016 false
2.017 bacteria
2.018 cholera
2.019 Louis Pasteur
2.020 smallpox
2.021 microscope
2.022 d
2.023 a
2.024 c
2.025 b
2.026 c

SELF TEST 3

3.01 c
3.02 d
3.03 b
3.04 h
3.05 j
3.06 f
3.07 i
3.08 e
3.09 a
3.010 false
3.011 true
3.012 true
3.013 true
3.014 h
3.015 k
3.016 r
3.017 l
3.018 a
3.019 b
3.020 m
3.021 n
3.022 i
3.023 f
3.024 o
3.025 j
3.026 d
3.027 p
3.028 c
3.029 g
3.030 e
3.031 d
3.032 a
3.033 a
3.034 c
3.035 b
3.036 Any order:
a. personal hygiene
b. living conditions
c. recreation
d. pollution
e. political factors
3.037 ✓
3.038 none
3.039 none
3.040 ✓
3.041 ✓
3.042 ✓
3.043 ✓
3.044 none
3.045 ✓
3.046 ✓

SELF TEST 1

1.01 d
1.02 a
1.03 f
1.04 b
1.05 g
1.06 Either order:
  a. Mercury
  b. Mars
  c. Venus
1.07 Any order:
  a. Mercury
  b. Mars
  c. Earth
  d. Jupiter
  e. Venus
  f. Saturn
1.08 18 trillion
1.09 3.3
1.010 3 millionths
1.011 2.5:1
1.012 1/3600
1.013 2
1.014 a. Cassiopeia
  b. Big Dipper (Ursa Major)
  c. Draco
  d. Cepheus
  e. Ursa Minor
  f. Ursa Major
  g. Camelopardalis
  h. Lynx

SELF TEST 2

2.01 false
2.02 true
2.03 true
2.04 false
2.05 true
2.06 d
2.07 a
2.08 f
2.09 c
2.010 b
2.011 g
2.012 i
2.013 2,500 times brighter
  or 1:2,500
2.014 660 inches
2.015 2400X
2.016 a. one-tenth
  b. 400
2.017 19
2.018 Pluto
2.019 Southern Hemisphere
2.020 light-year

SELF TEST 3

3.01 false
3.02 true
3.03 false
3.04 false
3.05 true
3.06 e
3.07 a
3.08 g
3.09 f
3.010 h
3.011 b
3.012 c
3.013 d
3.014 the basic building blocks of life
3.015 Either order:
  a. physiological
  b. biological
3.016 a. 1400
  b. 1730
3.017 2.8
3.018 Galileo

3.019 (1) raising it to the proper elevation above the earth's surface; (2) orienting it in the proper attitude; and (3) giving it the proper speed

3.020 the great circle on the celestial sphere that passes through the celestial poles and the observer's zenith

3.021 The astronomical unit is the distance of the earth from the sun or 93,000,000 miles.

3.022 Stars are large globes of intensely heated gas, which generate their own light. Nebulae are vast clouds of dust and gas made visible by the light of stars.

3.023 Skylab 4

3.024 Voyager

3.025 Pioneers 10 and 11

3.026 Sputnik

3.027 Apollo

SELF TEST 1

1.01 significant wave
1.02 Either order:
  a. *Nautilus*
  b. *Skate*
1.03 fractures
1.04 from the bottom
1.05 electromagnetic radiation
1.06 g
1.07 e
1.08 d
1.09 b
1.010 h
1.011 i
1.012 c
1.013 a
1.014 j
1.015 f
1.016 Either order:
  a. Edward Forbes
  b. Matthew Maury
1.017 Any order:
  a. Alexander Agassiz
  b. W.C. McIntosh
  c. Frank Buckland
1.018 a. Sealab I
  b. DSRV-1
  c. Trieste
  d. Sealab II

SELF TEST 2

2.01 false
2.02 true
2.03 false
2.04 true
2.05 true
2.06 Palomares, Spain
2.07 17,000 feet
2.08 Either order:
  a. mid-ocean ridges
  b. ocean trenches
2.09 Pacific Plate
2.010 turbidity currents
2.011 d
2.012 h
2.013 a
2.014 j
2.015 b
2.016 e
2.017 c
2.018 i
2.019 f
2.020 g
2.021 28.75 mph
2.022 62.5^3 in
2.023 100 million years

SELF TEST 3

3.01 This order only:
  a. Japan
  b. Soviet Union
  c. China (mainland)
  d. United States
  e. Chile
  f. Peru
3.02 Any order:
  a. amount of dissolved material
  b. chlorine content
  c. amount of evaporation
  d. amount of precipitation
  e. amount of fresh water introduced by rivers
  f. the mixing of currents
3.03 false
3.04 true
3.05 true
3.06 false
3.07 true
3.08 See illustration, page 24.
3.09 See illustration, page 26.

3.010 See illustration, page 27.
3.011 See illustration, page 16.
3.012 0.3 knots
3.013 f
3.014 j
3.015 l
3.016 a
3.017 b
3.018 m
3.019 c
3.020 d
3.021 e
3.022 g
3.023 h
3.024 i
3.025 k
3.026 n

SELF TEST 1

1.01 e
1.02 g
1.03 j
1.04 b
1.05 a
1.06 i
1.07 c
1.08 k
1.09 d
1.010 f
1.011 l
1.012 h
1.013 the dumping of wastes into rivers, lakes, and oceans
1.014 The thin top layer of land that provides plants with nutrients
1.015 the changing of water vapor to precipitation
1.016 true
1.017 false
1.018 false
1.019 true
1.020 true
1.021 false
1.022 true
1.023 Any order:
a. produce less
b. throw away less
c. recycle more efficiently
1.024 Any order:
a. unhealthy competition for necessities of life
b. insufficient medical care
c. poor effect on mental condition
1.025 Any order:
a. glass
b. rubber
c. cellulose
d. garbage
e. ash and cinder
1.026 a
1.027 d
1.028 b
1.029 c
1.030 b
1.031 c
1.032 a

SELF TEST 2

2.01 energy
2.02 protein
2.03 rapid transit system
2.04 hydrofoil
2.05 automobile
2.06 increasing
2.07 Any order:
a. oil
b. coal
c. natural gas
2.08 Any order:
a. produce less
b. throw away less
c. recycle more efficiently
2.09 Either order:
a. fission
b. fusion
2.010 c
2.011 h
2.012 g
2.013 i
2.014 k
2.015 b
2.016 l
2.017 d
2.018 j
2.019 a
2.020 e

2.021 the condition of becoming a city
2.022 the outlying areas of a city (i.e., population centers just beyond the city limits)
2.023 atomic energy that is released when atoms are split
2.024 true
2.025 true
2.026 true
2.027 false
2.028 true
2.029 false
2.030 false

SELF TEST 3

3.01 true
3.02 true
3.03 false
3.04 true
3.05 true
3.06 false
3.07 true
3.08 false
3.09 true
3.010 true
3.011 i
3.012 b
3.013 j
3.014 a
3.015 h
3.016 d
3.017 e
3.018 c
3.019 f
3.020 g
3.021 a person who travels beyond the earth's atmosphere; generally American
3.022 the sensing and transmitting of information about someone or something at great distances
3.023 the part of the ocean closest to land, averaging 600-800 feet deep, and containing 50% of all ocean life
3.024 Either order:
a. helium
b. oxygen
3.025 Any order:
a. air
b. water
c. waste
3.026 Genesis
3.027 Either order:
a. excessive fishing
b. encroachment
3.028 thermal
3.029 b
3.030 c
3.031 a
3.032 b
3.033 b
3.034 c
3.035 a
3.036 d
3.037 d
3.038 a

SELF TEST 1

1.01 .065
1.02 .28
1.03 21,000
1.04 8.32
1.05 4,500
1.06 35
1.07 710
1.08 .0061
1.09 3
1.010 10,5000
1.011 .032
1.012 1.5
1.013 1,200
1.014 350
1.015 6.25
1.016 .5
1.017 metric
1.018 the United States
1.019 weight
1.020 the amount of matter in an object
1.021 the measure of the pull of gravity on an object
1.022 4.8
1.023 1.9
1.024 38
1.025 54
1.026 15.24
1.027 On the planet's surface mass is the same as on the earth but weight would vary with each planet's gravitational pull.
1.028 Mass is still the same as on the earth but lack of gravity would cause total weightlessness.

SELF TEST 2

2.01 g
2.02 k
2.03 j
2.04 a
2.05 c
2.06 e
2.07 i
2.08 h
2.09 b
2.010 d
2.011 a. burying
b. water supplies
2.012 smallpox
2.013 Either order:
a. yellow fever
b. malaria
2.014 Either order:
a. rats
b. fleas
2.015 Either order:
a. you do not become ill
b. you cannot carry the disease
2.016 a. liter
b. meter
c. kilogram
2.017 Any order:
a. skin
b. secretions
c. white blood cells or leukocytes
d. antibodies
2.018 the United States
2.019 smallpox
2.020 Any order:
a. cholera
b. amoebic dysentery
c. bacillary dysentery
2.021 false
2.022 true

2.023 false
2.024 true
2.025 true
2.026 immunizations, drink only potable water, avoid unsanitary conditions, and so forth.
2.027 Our body is the temple of God. We must glorify God in our body and spirit.

SELF TEST 3

3.01 false
3.02 true
3.03 false
3.04 false
3.05 true
3.06 false
3.07 true
3.08 true
3.09 true
3.010 true
3.011 g
3.012 i
3.013 e
3.014 j
3.015 a
3.016 c
3.017 k
3.018 b
3.019 d
3.020 f
3.021 Any order:
  a. wind
  b. water
  c. glaciers
3.022 weight
3.023 schistosomiasis
3.024 Either order:
  a. malaria
  b. yellow fever
3.025 ellipse
3.026 Either order:
  a. midocean ridges
  b. trenches
3.027 folding
3.028 the San Andreas
3.029 plate tectonics
3.030 decimals (powers of ten)
3.031 Any order:
  a. volcanoes
  b. faults
  c. folds
  d. sedimentation
3.032 Any order:
  a. fossils
  b. lava flows
  c. rock layers
  d. glacial deposits
3.033 Mass is the amount of matter an object has. Weight is the pull of gravity on that object.
3.034 erosion of existing mountains

SELF TEST 4

4.01 i
4.02 k
4.03 h
4.04 j
4.05 a
4.06 c
4.07 f
4.08 b
4.09 g
4.010 d
4.011 a
4.012 d
4.013 b
4.014 d

4.015 d
4.016 a
4.017 d
4.018 a
4.019 c
4.020 d
4.021 c
4.022 f
4.023 h
4.024 k
4.025 j
4.026 b
4.027 g
4.028 i
4.029 a
4.030 e
4.031 Any order:
   a. atomic energy
   b. wind
   c. solar energy
   d. hydroelectric power
   e. thermal power
4.032 Any order:
   a. electrons
   b. proton
   c. neutron

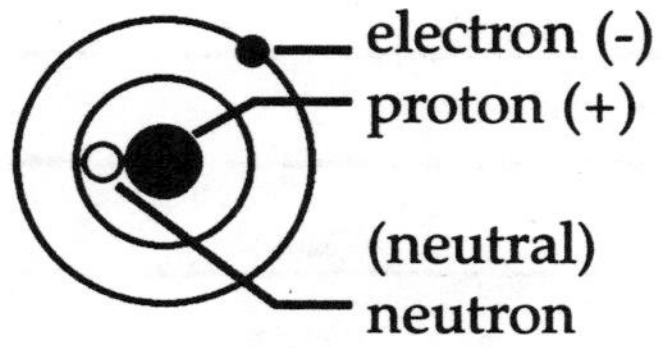

electron (-)
proton (+)
(neutral)
neutron

4.033 The joining together of light elements resulting in a heavier element plus energy
4.034 the splitting of a heavy element into lighter elements plus energy

Notes

1. false
2. true
3. false
4. false
5. true
6. true
7. false
8. true
9. true
10. true
11. c
12. d
13. f
14. a
15. b
16. a
17. b
18. c
19. c
20. a
21. a
22. c
23. a
24. b
25. c
26. a. energy or speed, distance, direction
    b. intensity
27. fermium
28. a. underground tanks
    b. salt mines
29. Any order:
    a. mesons
    b. positrons
    c. neutrinos
30. Marie and Pierre Curie
31. Any two; any order:
    a. irrigation of crops, prolonged shipping season,
    b. increased growth in ocean
32. a. protons
    b. neutrons
    c. electrons
33. Any order:
    a. curie
    b. roentgen
34. critical mass
35. radiation biological material absorbs.
36. inserting a substance that will absorb or buffer the decay products of unstable nuclei.
37. electron

1. c
2. b
3. c
4. a
5. a
6. false
7. true
8. true
9. false
10. false
11. true
12. liter
13. kilogram
14. gravity (weight)
15. direct
16. density
17. closeness to a standard
18. consistency of result
19. that quantity against which others are measured
20. mass per unit volume
21. density $= \frac{\text{mass}}{\text{volume}} = \frac{200\text{ g}}{40\text{ cm}^3} = 5\text{ g/cm}^3$

    Specific gravity is numerically equal to density, but without units.
    Specific gravity = 5

1. c
2. e
3. a
4. b
5. d
6. c
7. b
8. a
9. a
10. c
11. c
12. b
13. c
14. a
15. a
16. d
17. f
18. a
19. b
20. c
21. false
22. true
23. false
24. true
25. false
26. true
27. true
28. true
29. false
30. false
31. true
32. true
33. true
34. false
35. true
36. false
37. false
38. false
39. false
40. false
41. c
42. d
43. a
44. b
45. c
46. a
47. c
48. c
49. a
50. b

1. graded bed
2. chemical
3. mud
4. conglomerate
5. lithification
6. recrystallization
7. petrifaction
8. orogeny
9. paleontology
10. underwater
11. unconformity
12. thrust
13. Superposition
14. absolute
15. varves
16. c
17. a
18. c
19. a
20. c
21. a. group
    b. formation
    c. member
    d. bed
22. clay
23. silt
24. sand
25. sponge spines/reef frogs/shell fragments/ shells
26. gravel

1. false
2. true
3. true
4. false
5. true
6. false
7. false
8. true
9. true
10. true
11. true
12. false
13. true
14. true
15. false
16. true
17. true
18. true
19. true
20. true
21. true
22. true
23. true
24. true
25. false
26. c
27. a
28. d
29. b
30. h
31. g
32. j
33. i
34. f
35. d
36. c
37. a
38. b
39. d
40. c
41. a
42. b
43. c
44. d

Science 906 Test Key

1. d
2. f
3. a
4. b
5. e
6. c
7. j
8. h
9. i
10. k
11. b
12. d
13. c
14. a
15. b
16. c
17. a
18. d
19. c
20. a
21. c
22. b
23. c
24. a
25. d

1. satellite
2. Either order:
   a. Venus
   b. Mercury
3. 1 million (1,000,000)
4. Alpha-Centauri
5. Neptune
6. parallax
7. sixth
8. refracting
9. blue
10. increase the focal length
11. focal length
12. achromatic refractors
13. silver
14. 236 inches
15. Pioneer
16. SETI
17. 100 years
18. d
19. b
20. a
21. c
22. f
23. g
24. e
25. 500 A.U. = 5×10^2 A.U. x $93 \times 10^6 \frac{\text{mile}}{\text{A.U.}}$

   = 4.65×10^{10} mile

   1 LY = 6×10^{12} mile therefore,

   $$500 \text{ A.U.} = \frac{4.65 \times 10^{10} \text{ mile}}{6 \times 10^{12} \text{ mile/LY}}$$
   $$= 0.78 \times 10^{-2} \text{ LY}$$
   $$= 7.8 \times 10^{-3} \text{ LY}$$

26. c
27. f
28. a
29. e
30. b

1. b
2. a
3. c
4. b
5. a
6. true
7. false
8. true
9. false
10. false
11. true
12. false
13. true
14. true
15. false
16. a unit of measure equal to six feet, used mostly in measuring depth in water
17. the process by which simple sugars and starches are produced from carbon dioxide and water by living plant cells with the aid of chlorophyll and in the presence of light
18. small animal and plant organisms that float or drift in water which serve as an important source of food for larger fish
19. the process used to produce fresh water from saline, usually by distillation, but also to a small degree by crystallization
20. a pattern of feeding relationships among organism
21. Victor Hansen
22. C.G.J. Petersen
23. Cousteau and Gagnan
24. seismic refraction
25. Any order:
    a. subterranean earthquake
    b. turbidity currents
    c. volcanic eruptions
26. DSRV-1
27. clockwise
28. 8 feet

1. b
2. g
3. k
4. l
5. j
6. f
7. m
8. c
9. e
10. h
11. i
12. d
13. Any order:
    a. fossil fuels (oil, coal, natural gas)
    b. nuclear
    c. natural (wind, water, solar)
14. Any order:
    a. produce less
    b. throw away less
    c. recycle more efficiently
15. Any order:
    a. air
    b. land
    c. water
16. false
17. true
18. true
19. false
20. true
21. false
22. true
23. a
24. d
25. b
26. a
27. c
28. c
29. b
30. a
31. a. F
    b. A
    c. N
    d. F
    e. A
    f. F
    g. N
    h. N

Science 910 Test Key

1. true
2. true
3. true
4. true
5. false
6. true
7. true
8. true
9. true
10. true
11. h
12. e
13. g
14. b
15. c
16. a
17. d
18. i
19. j
20. k
21. weight
22. erosion
23. vaccination
24. improper food handling at picnics
25. midocean ridge
26. nucleus
27. sedimentary rock
28. smallpox
29. fusion
30. solar system
31. 600
32. .7
33. .05
34. 12,000
35. 400
36. 4.582

Table of Contents

Student Name: ____________________ Notebook Number: ____________________

Email: ____________________ Phone: ____________________

Network ID: ____________________ Course: ____________________

Lab Instructor: ____________________ Section: __________ Semester: __________

Lab Partners: ____

Date	Experiment/Subject	Page Number

Table of Contents

Date	Experiment/Subject	Page Number

Laboratory Safety

Laboratory safety should be the number one priority for you, your fellow students, and your instructors. The guidelines seen below are general laboratory safety practices that can apply to most laboratory settings. Your instructor will advise you of any specific precautions or hazards you may deal with in a particular experiment. Do not hesitate to ask your instructor about any questions you might have about an experiment or any of the safety precautions outlined here or given to you by your instructor.

- Never work alone in the lab without supervision and work only on an experiment that has been assigned to you.
- Be prepared for your lab by reading the experiment ahead of time and completing any pre-lab activities if there are any.
- For many laboratory experiments, safety eyewear should be worn at all times. Consult your instructor about the need for and kind of safety eyewear, such as goggles, and determine the policy for wearing contact lenses in the laboratory.
- Dress appropriately for lab, with clothing that covers your torso and legs and shoes that cover your entire foot. Tie back loose hair and avoid wearing baggy sleeves or dangling jewelry. Your instructor will inform you if you are to wear a lab coat or lab apron.
- For some experiments you may be required to wear gloves. Be sure to remove the gloves when leaving the lab even if you are planning to return right away. In addition, always wash your hands before leaving the lab.
- Eating, drinking, applying makeup, etc., are forbidden in the laboratory.

- Pipetting by mouth is never allowed in the laboratory.
- Dispose of all chemical waste or other waste that is produced during the lab in an appropriate manner designated by your instructor.
- Be sure you know the location of all safety equipment such as the safety shower, eyewash fountains, fire extinguishers, and emergency telephones. Report any accident, breakage of glassware, or spills to your instructor immediately.
- It is important to know how to evacuate from the laboratory in the event of a fire or other emergency.

These general safety guidelines are designed to help keep your laboratory environment as safe as it can be. It is important that you and the other students in the lab act responsibly and be diligent in following all safety rules outlined here and given to you by your instructor. Please sign below to acknowledge that you have read the guidelines and agree to follow them and any other directives given to you by your instructor.

______________________________ ________________

Student Signature Date

Laboratory Safety

Laboratory safety should be the number one priority for you, your fellow students, and your instructors. The guidelines seen below are general laboratory safety practices that can apply to most laboratory settings. Your instructor will advise you of any specific precautions or hazards you may deal with in a particular experiment. Do not hesitate to ask your instructor about any questions you might have about an experiment or any of the safety precautions outlined here or given to you by your instructor.

- Never work alone in the lab without supervision and work only on an experiment that has been assigned to you.
- Be prepared for your lab by reading the experiment ahead of time and completing any pre-lab activities if there are any.
- For many laboratory experiments, safety eyewear should be worn at all times. Consult your instructor about the need for and kind of safety eyewear, such as goggles, and determine the policy for wearing contact lenses in the laboratory.
- Dress appropriately for lab, with clothing that covers your torso and legs and shoes that cover your entire foot. Tie back loose hair and avoid wearing baggy sleeves or dangling jewelry. Your instructor will inform you if you are to wear a lab coat or lab apron.
- For some experiments you may be required to wear gloves. Be sure to remove the gloves when leaving the lab even if you are planning to return right away. In addition, always wash your hands before leaving the lab.
- Eating, drinking, applying makeup, etc., are forbidden in the laboratory.

- Pipetting by mouth is never allowed in the laboratory.
- Dispose of all chemical waste or other waste that is produced during the lab in an appropriate manner designated by your instructor.
- Be sure you know the location of all safety equipment such as the safety shower, eyewash fountains, fire extinguishers, and emergency telephones. Report any accident, breakage of glassware, or spills to your instructor immediately.
- It is important to know how to evacuate from the laboratory in the event of a fire or other emergency.

These general safety guidelines are designed to help keep your laboratory environment as safe as it can be. It is important that you and the other students in the lab act responsibly and be diligent in following all safety rules outlined here and given to you by your instructor. Please sign below to acknowledge that you have read the guidelines and agree to follow them and any other directives given to you by your instructor.

Student Signature ______________________ Date ____________

Exp. No.	Experiment/Subject	Date	
Name	Lab Partner	Locker/ Desk No.	Course & Section No.

Signature	Date	Witness/TA	Date

Note: Insert Divider Under Copy Sheet Before Writing

Exp. No.	Experiment/Subject		Date	
Name	Lab Partner		Locker/ Desk No.	Course & Section No.

Signature	Date	Witness/TA	Date

Exp. No.	Experiment/Subject	Date

Name	Lab Partner	Locker/ Desk No.	Course & Section No.

Signature	Date	Witness/TA	Date

02

Exp. No.	Experiment/Subject	Date	
Name	Lab Partner	Locker/ Desk No.	Course & Section No.

Signature	Date	Witness/TA	Date

Note: Insert Divider Under Copy Sheet Before Writing

Exp. No.	Experiment/Subject	Date	
Name	Lab Partner	Locker/ Desk No.	Course & Section No.

Signature	Date	Witness/TA	Date

Note: Insert Divider Under Copy Sheet Before Writing

Exp. No.	Experiment/Subject	Date	
Name	Lab Partner	Locker/ Desk No.	Course & Section No.

Signature	Date	Witness/TA	Date

Exp. No.	Experiment/Subject		Date	
Name	Lab Partner		Locker/ Desk No.	Course & Section No.

Signature	Date	Witness/TA	Date

Exp. No.	Experiment/Subject	Date	
Name	Lab Partner	Locker/ Desk No.	Course & Section No.

Signature	Date	Witness/TA	Date

Note: Insert Divider Under Copy Sheet Before Writing

Exp. No.	Experiment/Subject	Date	
Name	Lab Partner	Locker/ Desk No.	Course & Section No.

Signature	Date	Witness/TA	Date

Note: Insert Divider Under Copy Sheet Before Writing

Exp. No.	Experiment/Subject	Date	
Name	Lab Partner	Locker/ Desk No.	Course & Section No.

Signature	Date	Witness/TA	Date

Note: Insert Divider Under Copy Sheet Before Writing

Exp. No.	Experiment/Subject		Date	
Name		Lab Partner	Locker/ Desk No.	Course & Section No.

Signature	Date	Witness/TA	Date

Note: Insert Divider Under Copy Sheet Before Writing

Exp. No.	Experiment/Subject		Date	
Name		Lab Partner	Locker/ Desk No.	Course & Section No.

Signature	Date	Witness/TA	Date

Note: Insert Divider Under Copy Sheet Before Writing

Exp. No.	Experiment/Subject	Date	
Name	Lab Partner	Locker/ Desk No.	Course & Section No.

Signature	Date	Witness/TA	Date

Note: Insert Divider Under Copy Sheet Before Writing

Exp. No.	Experiment/Subject	Date	
Name	Lab Partner	Locker/ Desk No.	Course & Section No.

Signature	Date	Witness/TA	Date

Exp. No.	Experiment/Subject		Date	
Name		Lab Partner	Locker/ Desk No.	Course & Section No.

Signature	Date	Witness/TA	Date

Exp. No.	Experiment/Subject	Date	
Name	Lab Partner	Locker/ Desk No.	Course & Section No.

Signature	Date	Witness/TA	Date

Note: Insert Divider Under Copy Sheet Before Writing

Exp. No.	Experiment/Subject	Date

Name	Lab Partner	Locker/ Desk No.	Course & Section No.

Signature	Date	Witness/TA	Date

Exp. No.	Experiment/Subject	Date

Name	Lab Partner	Locker/ Desk No.	Course & Section No.

Signature	Date	Witness/TA	Date

Exp. No.	Experiment/Subject		Date	
Name		Lab Partner	Locker/ Desk No.	Course & Section No.

Signature	Date	Witness/TA	Date

Note: Insert Divider Under Copy Sheet Before Writing

Exp. No.	Experiment/Subject		Date	
Name		Lab Partner	Locker/ Desk No.	Course & Section No.

Signature	Date	Witness/TA	Date

Note: Insert Divider Under Copy Sheet Before Writing

Exp. No.	Experiment/Subject	Date	
Name	Lab Partner	Locker/ Desk No.	Course & Section No.

Signature	Date	Witness/TA	Date

Note: Insert Divider Under Copy Sheet Before Writing

Exp. No.	Experiment/Subject		Date	
Name		Lab Partner	Locker/ Desk No.	Course & Section No.

Signature	Date	Witness/TA	Date

Note: Insert Divider Under Copy Sheet Before Writing

Exp. No.	Experiment/Subject	Date	
Name	**Lab Partner**	**Locker/ Desk No.**	**Course & Section No.**

Signature	Date	Witness/TA	Date

Note: Insert Divider Under Copy Sheet Before Writing

Exp. No.	Experiment/Subject	Date	
Name	Lab Partner	Locker/ Desk No.	Course & Section No.

Signature	Date	Witness/TA	Date

Exp. No.	Experiment/Subject		Date	
Name		Lab Partner	Locker/ Desk No.	Course & Section No.

Signature	Date	Witness/TA	Date

Exp. No.	Experiment/Subject		Date	
Name		Lab Partner	Locker/ Desk No.	Course & Section No.

Signature	Date	Witness/TA	Date

Exp. No.	Experiment/Subject	Date	
Name	Lab Partner	Locker/ Desk No.	Course & Section No.

Signature	Date	Witness/TA	Date

Note: Insert Divider Under Copy Sheet Before Writing

Exp. No.	Experiment/Subject	Date	
Name	Lab Partner	Locker/ Desk No.	Course & Section No.

Signature	Date	Witness/TA	Date

Exp. No.	Experiment/Subject		Date	
Name		Lab Partner	Locker/ Desk No.	Course & Section No.

Signature	Date	Witness/TA	Date

Note: Insert Divider Under Copy Sheet Before Writing

Exp. No.	Experiment/Subject	Date	
Name	**Lab Partner**	**Locker/ Desk No.**	**Course & Section No.**

Signature	Date	Witness/TA	Date

Exp. No.	Experiment/Subject	Date	
Name	Lab Partner	Locker/ Desk No.	Course & Section No.

Signature	Date	Witness/TA	Date

Note: Insert Divider Under Copy Sheet Before Writing

Exp. No.	Experiment/Subject	Date	
Name	**Lab Partner**	**Locker/ Desk No.**	**Course & Section No.**

Signature	Date	Witness/TA	Date

Exp. No.	Experiment/Subject		Date	
Name		Lab Partner	Locker/ Desk No.	Course & Section No.

Signature	Date	Witness/TA	Date

Note: Insert Divider Under Copy Sheet Before Writing

Exp. No.	Experiment/Subject	Date	
Name	Lab Partner	Locker/ Desk No.	Course & Section No.

COPY

Signature	Date	Witness/TA	Date

Note: Insert Divider Under Copy Sheet Before Writing

Exp. No.	Experiment/Subject		Date	
Name	Lab Partner		Locker/ Desk No.	Course & Section No.

Signature	Date	Witness/TA	Date

Note: Insert Divider Under Copy Sheet Before Writing

Exp. No.	Experiment/Subject	Date	
Name	Lab Partner	Locker/ Desk No.	Course & Section No.

Signature	Date	Witness/TA	Date

Exp. No.	Experiment/Subject	Date	
Name	Lab Partner	Locker/ Desk No.	Course & Section No.

Signature	Date	Witness/TA	Date

Note: Insert Divider Under Copy Sheet Before Writing

Exp. No.	Experiment/Subject	Date	
Name	Lab Partner	Locker/ Desk No.	Course & Section No.

Signature	Date	Witness/TA	Date

Note: Insert Divider Under Copy Sheet Before Writing

Exp. No.	Experiment/Subject	Date	
Name	Lab Partner	Locker/ Desk No.	Course & Section No.

Signature	Date	Witness/TA	Date

Note: Insert Divider Under Copy Sheet Before Writing

Exp. No.	Experiment/Subject	Date	
Name	Lab Partner	Locker/ Desk No.	Course & Section No.

Signature	Date	Witness/TA	Date

Exp. No.	Experiment/Subject	Date	
Name	Lab Partner	Locker/ Desk No.	Course & Section No.

Signature	Date	Witness/TA	Date

Note: Insert Divider Under Copy Sheet Before Writing

Exp. No.	Experiment/Subject	Date	
Name	Lab Partner	Locker/ Desk No.	Course & Section No.

Signature	Date	Witness/TA	Date

Exp. No.	Experiment/Subject		Date	
Name		Lab Partner	Locker/ Desk No.	Course & Section No.

Signature	Date	Witness/TA	Date

Note: Insert Divider Under Copy Sheet Before Writing

Exp. No.	Experiment/Subject	Date	
Name	Lab Partner	Locker/ Desk No.	Course & Section No.

Signature	Date	Witness/TA	Date

Note: Insert Divider Under Copy Sheet Before Writing

Exp. No.	Experiment/Subject	Date	
Name	Lab Partner	Locker/ Desk No.	Course & Section No.

Signature	Date	Witness/TA	Date

Note: Insert Divider Under Copy Sheet Before Writing

Exp. No.	Experiment/Subject	Date	
Name	Lab Partner	Locker/ Desk No.	Course & Section No.

Signature	Date	Witness/TA	Date

Exp. No.	Experiment/Subject		Date	
Name	Lab Partner		Locker/ Desk No.	Course & Section No.

Signature	Date	Witness/TA	Date

Note: Insert Divider Under Copy Sheet Before Writing

Exp. No.	Experiment/Subject	Date	
Name	Lab Partner	Locker/ Desk No.	Course & Section No.

Signature	Date	Witness/TA	Date

Exp. No.	Experiment/Subject		Date	
Name		Lab Partner	Locker/ Desk No.	Course & Section No.

Signature	Date	Witness/TA	Date

Exp. No.	Experiment/Subject	Date	
Name	Lab Partner	Locker/ Desk No.	Course & Section No.

Signature	Date	Witness/TA	Date

Note: Insert Divider Under Copy Sheet Before Writing

Exp. No.	Experiment/Subject	Date	
Name	Lab Partner	Locker/ Desk No.	Course & Section No.

Signature	Date	Witness/TA	Date

Exp. No.	Experiment/Subject	Date	
Name	Lab Partner	Locker/ Desk No.	Course & Section No.

Signature	Date	Witness/TA	Date

Exp. No.	Experiment/Subject	Date	
Name	Lab Partner	Locker/ Desk No.	Course & Section No.

Signature	Date	Witness/TA	Date

Note: Insert Divider Under Copy Sheet Before Writing

Exp. No.	Experiment/Subject	Date	
Name	Lab Partner	Locker/ Desk No.	Course & Section No.

Signature	Date	Witness/TA	Date

Exp. No.	Experiment/Subject	Date	
Name	Lab Partner	Locker/ Desk No.	Course & Section No.

Signature	Date	Witness/TA	Date

Note: Insert Divider Under Copy Sheet Before Writing

Exp. No.	Experiment/Subject		Date	
Name		Lab Partner	Locker/ Desk No.	Course & Section No.

Signature	Date	Witness/TA	Date

Note: Insert Divider Under Copy Sheet Before Writing

Exp. No.	Experiment/Subject		Date	
Name		Lab Partner	Locker/ Desk No.	Course & Section No.

Signature	Date	Witness/TA	Date

Note: Insert Divider Under Copy Sheet Before Writing

Exp. No.	Experiment/Subject	Date	
Name	Lab Partner	Locker/ Desk No.	Course & Section No.

Signature	Date	Witness/TA	Date

Exp. No.	Experiment/Subject	Date	
Name	Lab Partner	Locker/ Desk No.	Course & Section No.

Signature	Date	Witness/TA	Date

Note: Insert Divider Under Copy Sheet Before Writing

Exp. No.	Experiment/Subject	Date	
Name	Lab Partner	Locker/ Desk No.	Course & Section No.

Signature	Date	Witness/TA	Date

Exp. No.	Experiment/Subject		Date	
Name		Lab Partner	Locker/ Desk No.	Course & Section No.

Signature	Date	Witness/TA	Date

Exp. No.	Experiment/Subject		Date	
Name		Lab Partner	Locker/ Desk No.	Course & Section No.

Signature	Date	Witness/TA	Date

Exp. No.	Experiment/Subject	Date	
Name	Lab Partner	Locker/ Desk No.	Course & Section No.

Signature	Date	Witness/TA	Date

Note: Insert Divider Under Copy Sheet Before Writing

Exp. No.	Experiment/Subject	Date	
Name	Lab Partner	Locker/ Desk No.	Course & Section No.

Signature	Date	Witness/TA	Date

Note: Insert Divider Under Copy Sheet Before Writing

Exp. No.	Experiment/Subject	Date	
Name	Lab Partner	Locker/ Desk No.	Course & Section No.

Signature	Date	Witness/TA	Date

Exp. No.	Experiment/Subject	Date	
Name	Lab Partner	Locker/ Desk No.	Course & Section No.

Signature	Date	Witness/TA	Date

Exp. No.	Experiment/Subject	Date	
Name	Lab Partner	Locker/ Desk No.	Course & Section No.

Signature	Date	Witness/TA	Date

Note: Insert Divider Under Copy Sheet Before Writing

Exp. No.	Experiment/Subject	Date	
Name	Lab Partner	Locker/ Desk No.	Course & Section No.

Signature	Date	Witness/TA	Date

Note: Insert Divider Under Copy Sheet Before Writing

Exp. No.	Experiment/Subject		Date	
Name		Lab Partner	Locker/ Desk No.	Course & Section No.

Signature	Date	Witness/TA	Date

Exp. No.	Experiment/Subject	Date	
Name	Lab Partner	Locker/ Desk No.	Course & Section No.

COPY

Signature	Date	Witness/TA	Date

Note: Insert Divider Under Copy Sheet Before Writing

Exp. No.	Experiment/Subject		Date	
Name		Lab Partner	Locker/ Desk No.	Course & Section No.

Signature	Date	Witness/TA	Date

Note: Insert Divider Under Copy Sheet Before Writing

Exp. No.	Experiment/Subject	Date	
Name	**Lab Partner**	**Locker/ Desk No.**	**Course & Section No.**

Signature	Date	Witness/TA	Date

Exp. No.	Experiment/Subject	Date	
Name	Lab Partner	Locker/ Desk No.	Course & Section No.

Signature	Date	Witness/TA	Date

Note: Insert Divider Under Copy Sheet Before Writing

Exp. No.	Experiment/Subject	Date	
Name	Lab Partner	Locker/ Desk No.	Course & Section No.

Signature	Date	Witness/TA	Date

Note: Insert Divider Under Copy Sheet Before Writing

Exp. No.	Experiment/Subject	Date	
Name	Lab Partner	Locker/ Desk No.	Course & Section No.

Signature	Date	Witness/TA	Date

Note: Insert Divider Under Copy Sheet Before Writing

Exp. No.	Experiment/Subject	Date

Name	Lab Partner	Locker/ Desk No.	Course & Section No.

Signature	Date	Witness/TA	Date

Exp. No.	Experiment/Subject		Date	
Name		Lab Partner	Locker/ Desk No.	Course & Section No.

Signature	Date	Witness/TA	Date

Exp. No.	Experiment/Subject	Date	
Name	Lab Partner	Locker/ Desk No.	Course & Section No.

Signature	Date	Witness/TA	Date

Exp. No.	Experiment/Subject	Date	
Name	Lab Partner	Locker/ Desk No.	Course & Section No.

Signature	Date	Witness/TA	Date

Note: Insert Divider Under Copy Sheet Before Writing

Exp. No.	Experiment/Subject	Date

Name	Lab Partner	Locker/ Desk No.	Course & Section No.

Signature	Date	Witness/TA	Date

Exp. No.	Experiment/Subject		Date	
Name	Lab Partner		Locker/ Desk No.	Course & Section No.

Signature	Date	Witness/TA	Date

Exp. No.	Experiment/Subject	Date

Name	Lab Partner	Locker/ Desk No.	Course & Section No.

Signature	Date	Witness/TA	Date

Note: Insert Divider Under Copy Sheet Before Writing

Exp. No.	Experiment/Subject		Date	
Name		Lab Partner	Locker/ Desk No.	Course & Section No.

Signature	Date	Witness/TA	Date

Exp. No.	Experiment/Subject	Date	
Name	Lab Partner	Locker/ Desk No.	Course & Section No.

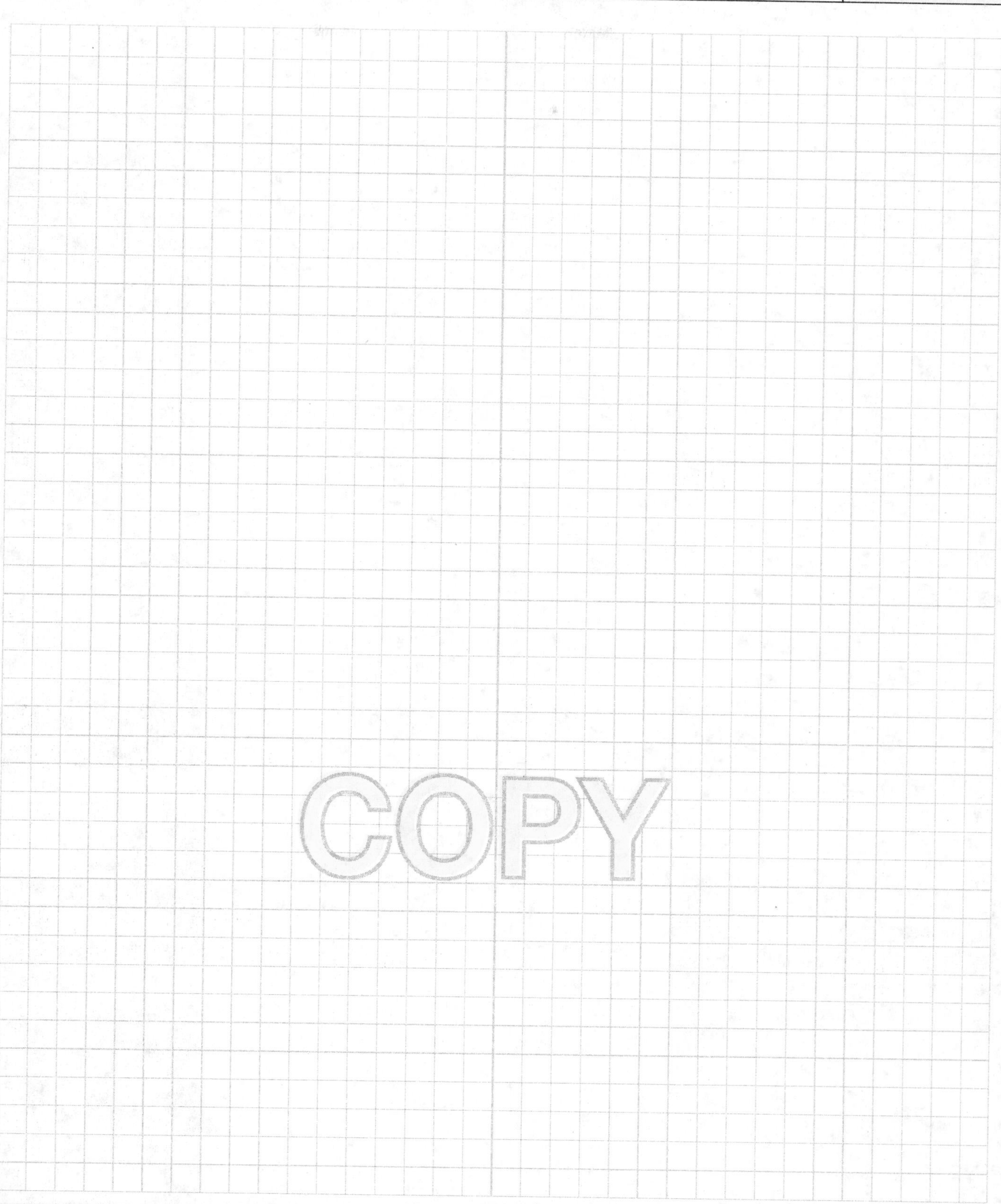

Signature	Date	Witness/TA	Date

Exp. No.	Experiment/Subject		Date	
Name		Lab Partner	Locker/ Desk No.	Course & Section No.

Signature	Date	Witness/TA	Date

Note: Insert Divider Under Copy Sheet Before Writing

Exp. No.	Experiment/Subject	Date	
Name	Lab Partner	Locker/ Desk No.	Course & Section No.

Signature	Date	Witness/TA	Date

Note: Insert Divider Under Copy Sheet Before Writing

Exp. No.	Experiment/Subject		Date	
Name		Lab Partner	Locker/ Desk No.	Course & Section No.

Signature	Date	Witness/TA	Date

Exp. No.	Experiment/Subject		Date	
Name		Lab Partner	Locker/ Desk No.	Course & Section No.

Signature	Date	Witness/TA	Date

Note: Insert Divider Under Copy Sheet Before Writing

Exp. No.	Experiment/Subject	Date	
Name	Lab Partner	Locker/ Desk No.	Course & Section No.

Signature	Date	Witness/TA	Date

Note: Insert Divider Under Copy Sheet Before Writing

Exp. No.	Experiment/Subject		Date	
Name		Lab Partner	Locker/ Desk No.	Course & Section No.

Signature	Date	Witness/TA	Date

Note: Insert Divider Under Copy Sheet Before Writing

Exp. No.	Experiment/Subject	Date	
Name	Lab Partner	Locker/ Desk No.	Course & Section No.

Signature	Date	Witness/TA	Date

Exp. No.	Experiment/Subject	Date	
Name	Lab Partner	Locker/ Desk No.	Course & Section No.

Signature	Date	Witness/TA	Date

Note: Insert Divider Under Copy Sheet Before Writing

Exp. No.	Experiment/Subject		Date	
Name		Lab Partner	Locker/ Desk No.	Course & Section No.

Signature	Date	Witness/TA	Date

Note: Insert Divider Under Copy Sheet Before Writing

Exp. No.	Experiment/Subject	Date	
Name	Lab Partner	Locker/ Desk No.	Course & Section No.

Signature	Date	Witness/TA	Date

Exp. No.	Experiment/Subject	Date	
Name	Lab Partner	Locker/ Desk No.	Course & Section No.

Signature	Date	Witness/TA	Date

Exp. No.	Experiment/Subject	Date	
Name	Lab Partner	Locker/ Desk No.	Course & Section No.

Signature	Date	Witness/TA	Date

Exp. No.	Experiment/Subject	Date	
Name	Lab Partner	Locker/ Desk No.	Course & Section No.

Signature	Date	Witness/TA	Date

Note: Insert Divider Under Copy Sheet Before Writing

Exp. No.	Experiment/Subject	Date	
Name	Lab Partner	Locker/ Desk No.	Course & Section No.

Signature	Date	Witness/TA	Date

Note: Insert Divider Under Copy Sheet Before Writing

Exp. No.	Experiment/Subject	Date

Name	Lab Partner	Locker/ Desk No.	Course & Section No.

Signature	Date	Witness/TA	Date

Note: Insert Divider Under Copy Sheet Before Writing

Exp. No.	Experiment/Subject	Date

Name	Lab Partner	Locker/ Desk No.	Course & Section No.

Signature	Date	Witness/TA	Date

Exp. No.	Experiment/Subject		Date	
Name		Lab Partner	Locker/ Desk No.	Course & Section No.

Signature	Date	Witness/TA	Date

Note: Insert Divider Under Copy Sheet Before Writing

Exp. No.	Experiment/Subject		Date
Name	Lab Partner	Locker/ Desk No.	Course & Section No.

Signature	Date	Witness/TA	Date

Exp. No.	Experiment/Subject		Date	
Name		Lab Partner	Locker/ Desk No.	Course & Section No.

Signature	Date	Witness/TA	Date

Note: Insert Divider Under Copy Sheet Before Writing

Exp. No.	Experiment/Subject		Date	
Name	Lab Partner		Locker/ Desk No.	Course & Section No.

Signature	Date	Witness/TA	Date

Exp. No.	Experiment/Subject	Date	
Name	Lab Partner	Locker/ Desk No.	Course & Section No.

Signature	Date	Witness/TA	Date

Note: Insert Divider Under Copy Sheet Before Writing

Exp. No.	Experiment/Subject		Date	
Name		Lab Partner	Locker/ Desk No.	Course & Section No.

Signature	Date	Witness/TA	Date

Exp. No.	Experiment/Subject		Date	
Name		Lab Partner	Locker/ Desk No.	Course & Section No.

Signature	Date	Witness/TA	Date

Exp. No.	Experiment/Subject		Date	
Name		Lab Partner	Locker/ Desk No.	Course & Section No.

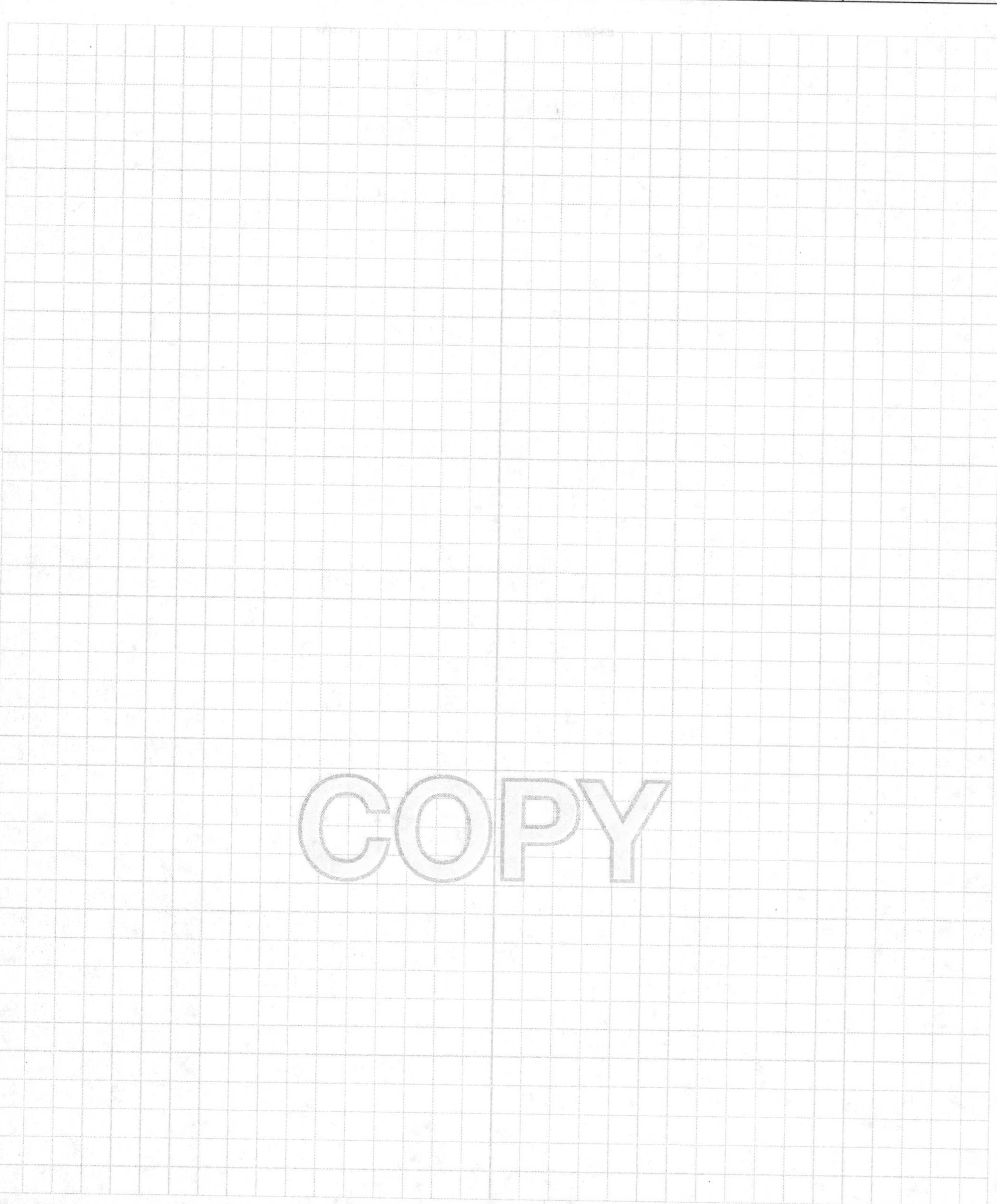

Signature	Date	Witness/TA	Date

Note: Insert Divider Under Copy Sheet Before Writing

Exp. No.	Experiment/Subject	Date	
Name	Lab Partner	Locker/ Desk No.	Course & Section No.

Signature	Date	Witness/TA	Date

Note: Insert Divider Under Copy Sheet Before Writing

Exp. No.	Experiment/Subject	Date	
Name	Lab Partner	Locker/ Desk No.	Course & Section No.

Signature	Date	Witness/TA	Date

Exp. No.	Experiment/Subject		Date	
Name	Lab Partner		Locker/ Desk No.	Course & Section No.

Signature	Date	Witness/TA	Date

Note: Insert Divider Under Copy Sheet Before Writing

Exp. No.	Experiment/Subject		Date	
Name		Lab Partner	Locker/ Desk No.	Course & Section No.

Signature	Date	Witness/TA	Date

Note: Insert Divider Under Copy Sheet Before Writing

Exp. No.	Experiment/Subject		Date	
Name		Lab Partner	Locker/ Desk No.	Course & Section No.

Signature	Date	Witness/TA	Date

Exp. No.	Experiment/Subject		Date	
Name		**Lab Partner**	**Locker/ Desk No.**	**Course & Section No.**

Signature	Date	Witness/TA	Date

Exp. No.	Experiment/Subject		Date	
Name		Lab Partner	Locker/ Desk No.	Course & Section No.

Signature	Date	Witness/TA	Date

Exp. No.	Experiment/Subject	Date	
Name	Lab Partner	Locker/ Desk No.	Course & Section No.

COPY

Signature	Date	Witness/TA	Date

Note: Insert Divider Under Copy Sheet Before Writing

Exp. No.	Experiment/Subject	Date	
Name	Lab Partner	Locker/ Desk No.	Course & Section No.

Signature	Date	Witness/TA	Date

Note: Insert Divider Under Copy Sheet Before Writing

Exp. No.	Experiment/Subject		Date	
Name		Lab Partner	Locker/ Desk No.	Course & Section No.

Signature	Date	Witness/TA	Date

Note: Insert Divider Under Copy Sheet Before Writing

Exp. No.	Experiment/Subject	Date	
Name	Lab Partner	Locker/ Desk No.	Course & Section No.

Signature	Date	Witness/TA	Date

Note: Insert Divider Under Copy Sheet Before Writing

Exp. No.	Experiment/Subject		Date	
Name		Lab Partner	Locker/ Desk No.	Course & Section No.

Signature	Date	Witness/TA	Date

Note: Insert Divider Under Copy Sheet Before Writing

Exp. No.	Experiment/Subject		Date	
Name		Lab Partner	Locker/ Desk No.	Course & Section No.

Signature	Date	Witness/TA	Date

Note: Insert Divider Under Copy Sheet Before Writing

Exp. No.	Experiment/Subject	Date	
Name	Lab Partner	Locker/ Desk No.	Course & Section No.

Signature	Date	Witness/TA	Date

Note: Insert Divider Under Copy Sheet Before Writing

Exp. No.	Experiment/Subject		Date	
Name		Lab Partner	Locker/ Desk No.	Course & Section No.

Signature	Date	Witness/TA	Date

Note: Insert Divider Under Copy Sheet Before Writing

Exp. No.	Experiment/Subject		Date	
Name		Lab Partner	Locker/ Desk No.	Course & Section No.

Signature	Date	Witness/TA	Date

Note: Insert Divider Under Copy Sheet Before Writing

Exp. No.	Experiment/Subject	Date	
Name	Lab Partner	Locker/ Desk No.	Course & Section No.

Signature	Date	Witness/TA	Date

Exp. No.	Experiment/Subject		Date	
Name		Lab Partner	Locker/ Desk No.	Course & Section No.

Signature	Date	Witness/TA	Date

Exp. No.	Experiment/Subject	Date	
Name	**Lab Partner**	**Locker/ Desk No.**	**Course & Section No.**

Signature	Date	Witness/TA	Date

Note: Insert Divider Under Copy Sheet Before Writing

Exp. No.	Experiment/Subject	Date	
Name	Lab Partner	Locker/ Desk No.	Course & Section No.

Signature	Date	Witness/TA	Date

Exp. No.	Experiment/Subject	Date	
Name	Lab Partner	Locker/ Desk No.	Course & Section No.

Signature	Date	Witness/TA	Date

Note: Insert Divider Under Copy Sheet Before Writing

Exp. No.	Experiment/Subject		Date	
Name		Lab Partner	Locker/ Desk No.	Course & Section No.

Signature	Date	Witness/TA	Date

Exp. No.	Experiment/Subject	Date	
Name	Lab Partner	Locker/ Desk No.	Course & Section No.

Signature	Date	Witness/TA	Date

Note: Insert Divider Under Copy Sheet Before Writing

Exp. No.	Experiment/Subject		Date	
Name		Lab Partner	Locker/ Desk No.	Course & Section No.

Signature	Date	Witness/TA	Date

Note: Insert Divider Under Copy Sheet Before Writing

Exp. No.	Experiment/Subject		Date	
Name		Lab Partner	Locker/ Desk No.	Course & Section No.

Signature	Date	Witness/TA	Date

Note: Insert Divider Under Copy Sheet Before Writing

Exp. No.	Experiment/Subject		Date	
Name		Lab Partner	Locker/ Desk No.	Course & Section No.

Signature	Date	Witness/TA	Date

Note: Insert Divider Under Copy Sheet Before Writing

Exp. No.	Experiment/Subject		Date	
Name	Lab Partner		Locker/ Desk No.	Course & Section No.

Signature	Date	Witness/TA	Date

Note: Insert Divider Under Copy Sheet Before Writing

Exp. No.	Experiment/Subject		Date	
Name		Lab Partner	Locker/ Desk No.	Course & Section No.

Signature	Date	Witness/TA	Date

Note: Insert Divider Under Copy Sheet Before Writing

Exp. No.	Experiment/Subject	Date	
Name	Lab Partner	Locker/ Desk No.	Course & Section No.

Signature	Date	Witness/TA	Date

Note: Insert Divider Under Copy Sheet Before Writing

Exp. No.	Experiment/Subject		Date	
Name	Lab Partner		Locker/ Desk No.	Course & Section No.

Signature	Date	Witness/TA	Date

Exp. No.	Experiment/Subject	Date	
Name	Lab Partner	Locker/ Desk No.	Course & Section No.

Signature	Date	Witness/TA	Date

Note: Insert Divider Under Copy Sheet Before Writing

Exp. No.	Experiment/Subject		Date	
Name		Lab Partner	Locker/ Desk No.	Course & Section No.

Signature	Date	Witness/TA	Date

Note: Insert Divider Under Copy Sheet Before Writing

Exp. No.	Experiment/Subject	Date	
Name	Lab Partner	Locker/ Desk No.	Course & Section No.

Signature	Date	Witness/TA	Date

Note: Insert Divider Under Copy Sheet Before Writing

Exp. No.	Experiment/Subject		Date	
Name		Lab Partner	Locker/ Desk No.	Course & Section No.

Signature	Date	Witness/TA	Date

Note: Insert Divider Under Copy Sheet Before Writing

Exp. No.	Experiment/Subject		Date	
Name	Lab Partner		Locker/ Desk No.	Course & Section No.

Signature	Date	Witness/TA	Date

Exp. No.	Experiment/Subject		Date	
Name		Lab Partner	Locker/ Desk No.	Course & Section No.

Signature	Date	Witness/TA	Date

Note: Insert Divider Under Copy Sheet Before Writing

Exp. No.	Experiment/Subject	Date	
Name	Lab Partner	Locker/ Desk No.	Course & Section No.

Signature	Date	Witness/TA	Date

Exp. No.	Experiment/Subject	Date	
Name	Lab Partner	Locker/ Desk No.	Course & Section No.

Signature	Date	Witness/TA	Date

Exp. No.	Experiment/Subject		Date	
Name		Lab Partner	Locker/ Desk No.	Course & Section No.

Signature	Date	Witness/TA	Date

Note: Insert Divider Under Copy Sheet Before Writing

Exp. No.	Experiment/Subject	Date	
Name	Lab Partner	Locker/ Desk No.	Course & Section No.

Signature	Date	Witness/TA	Date

Exp. No.	Experiment/Subject		Date	
Name		Lab Partner	Locker/ Desk No.	Course & Section No.

Signature	Date	Witness/TA	Date

Note: Insert Divider Under Copy Sheet Before Writing

Exp. No.	Experiment/Subject		Date	
Name		Lab Partner	Locker/ Desk No.	Course & Section No.

Signature	Date	Witness/TA	Date

Note: Insert Divider Under Copy Sheet Before Writing

Exp. No.	Experiment/Subject		Date	
Name		Lab Partner	Locker/ Desk No.	Course & Section No.

Signature	Date	Witness/TA	Date

Note: Insert Divider Under Copy Sheet Before Writing

Exp. No.	Experiment/Subject	Date	
Name	Lab Partner	Locker/ Desk No.	Course & Section No.

Signature	Date	Witness/TA	Date

Exp. No.	Experiment/Subject	Date	
Name	Lab Partner	Locker/ Desk No.	Course & Section No.

Signature	Date	Witness/TA	Date

Note: Insert Divider Under Copy Sheet Before Writing

Exp. No.	Experiment/Subject		Date	
Name		Lab Partner	Locker/ Desk No.	Course & Section No.

Signature	Date	Witness/TA	Date

Exp. No.	Experiment/Subject	Date	
Name	Lab Partner	Locker/ Desk No.	Course & Section No.

Signature	Date	Witness/TA	Date

Note: Insert Divider Under Copy Sheet Before Writing

Exp. No.	Experiment/Subject		Date	
Name		Lab Partner	Locker/ Desk No.	Course & Section No.

Signature	Date	Witness/TA	Date

Note: Insert Divider Under Copy Sheet Before Writing

Exp. No.	Experiment/Subject	Date	
Name	Lab Partner	Locker/ Desk No.	Course & Section No.

Signature	Date	Witness/TA	Date

Exp. No.	Experiment/Subject		Date	
Name		Lab Partner	Locker/ Desk No.	Course & Section No.

Signature	Date	Witness/TA	Date

Exp. No.	Experiment/Subject		Date	
Name		Lab Partner	Locker/ Desk No.	Course & Section No.

Signature	Date	Witness/TA	Date

Note: Insert Divider Under Copy Sheet Before Writing

Exp. No.	Experiment/Subject		Date	
Name		Lab Partner	Locker/ Desk No.	Course & Section No.

Signature	Date	Witness/TA	Date

Exp. No.	Experiment/Subject		Date	
Name		Lab Partner	Locker/ Desk No.	Course & Section No.

Signature	Date	Witness/TA	Date

Note: Insert Divider Under Copy Sheet Before Writing

Exp. No.	Experiment/Subject	Date	
Name	Lab Partner	Locker/ Desk No.	Course & Section No.

COPY

Signature	Date	Witness/TA	Date

Note: Insert Divider Under Copy Sheet Before Writing

Exp. No.	Experiment/Subject	Date	
Name	Lab Partner	Locker/ Desk No.	Course & Section No.

Signature	Date	Witness/TA	Date

Note: Insert Divider Under Copy Sheet Before Writing

Exp. No.	Experiment/Subject	Date	
Name	Lab Partner	Locker/ Desk No.	Course & Section No.

Signature	Date	Witness/TA	Date

Exp. No.	Experiment/Subject	Date	
Name	Lab Partner	Locker/ Desk No.	Course & Section No.

Signature	Date	Witness/TA	Date

Note: Insert Divider Under Copy Sheet Before Writing

Exp. No.	Experiment/Subject	Date	
Name	Lab Partner	Locker/ Desk No.	Course & Section No.

Signature	Date	Witness/TA	Date

Exp. No.	Experiment/Subject		Date	
Name		Lab Partner	Locker/ Desk No.	Course & Section No.

Signature	Date	Witness/TA	Date

Exp. No.	Experiment/Subject	Date	
Name	Lab Partner	Locker/ Desk No.	Course & Section No.

Signature	Date	Witness/TA	Date

Exp. No.	Experiment/Subject	Date	
Name	Lab Partner	Locker/ Desk No.	Course & Section No.

Signature	Date	Witness/TA	Date

Note: Insert Divider Under Copy Sheet Before Writing

Exp. No.	Experiment/Subject	Date	
Name	Lab Partner	Locker/ Desk No.	Course & Section No.

Signature	Date	Witness/TA	Date

Note: Insert Divider Under Copy Sheet Before Writing

Exp. No.	Experiment/Subject	Date	
Name	Lab Partner	Locker/ Desk No.	Course & Section No.

Signature	Date	Witness/TA	Date

Note: Insert Divider Under Copy Sheet Before Writing

Exp. No.	Experiment/Subject		Date	
Name	Lab Partner		Locker/ Desk No.	Course & Section No.

Signature	Date	Witness/TA	Date

Exp. No.	Experiment/Subject		Date	
Name		Lab Partner	Locker/ Desk No.	Course & Section No.

Signature	Date	Witness/TA	Date

Note: Insert Divider Under Copy Sheet Before Writing

Exp. No.	Experiment/Subject	Date	
Name	Lab Partner	Locker/ Desk No.	Course & Section No.

Signature	Date	Witness/TA	Date

Note: Insert Divider Under Copy Sheet Before Writing

Exp. No.	Experiment/Subject	Date	
Name	Lab Partner	Locker/ Desk No.	Course & Section No.

Signature	Date	Witness/TA	Date

Exp. No.	Experiment/Subject	Date	
Name	Lab Partner	Locker/ Desk No.	Course & Section No.

Signature	Date	Witness/TA	Date

Exp. No.	Experiment/Subject	Date	
Name	Lab Partner	Locker/ Desk No.	Course & Section No.

Signature	Date	Witness/TA	Date

Note: Insert Divider Under Copy Sheet Before Writing

Exp. No.	Experiment/Subject	Date	
Name	Lab Partner	Locker/ Desk No.	Course & Section No.

Signature	Date	Witness/TA	Date

Exp. No.	Experiment/Subject		Date	
Name		Lab Partner	Locker/ Desk No.	Course & Section No.

Signature	Date	Witness/TA	Date

Exp. No.	Experiment/Subject	Date	
Name	Lab Partner	Locker/ Desk No.	Course & Section No.

Signature	Date	Witness/TA	Date

Exp. No.	Experiment/Subject		Date	
Name		Lab Partner	Locker/ Desk No.	Course & Section No.

Signature	Date	Witness/TA	Date

Note: Insert Divider Under Copy Sheet Before Writing

Exp. No.	Experiment/Subject		Date	
Name		Lab Partner	Locker/ Desk No.	Course & Section No.

Signature	Date	Witness/TA	Date

Note: Insert Divider Under Copy Sheet Before Writing

Exp. No.	Experiment/Subject	Date	
Name	Lab Partner	Locker/ Desk No.	Course & Section No.

Signature	Date	Witness/TA	Date

Note: Insert Divider Under Copy Sheet Before Writing

Exp. No.	Experiment/Subject		Date	
Name		Lab Partner	Locker/ Desk No.	Course & Section No.

Signature	Date	Witness/TA	Date

Note: Insert Divider Under Copy Sheet Before Writing

Exp. No.	Experiment/Subject	Date	
Name	Lab Partner	Locker/ Desk No.	Course & Section No.

Signature	Date	Witness/TA	Date

Note: Insert Divider Under Copy Sheet Before Writing

Exp. No.	Experiment/Subject	Date	
Name	Lab Partner	Locker/ Desk No.	Course & Section No.

Signature	Date	Witness/TA	Date

Exp. No.	Experiment/Subject		Date	
Name		Lab Partner	Locker/ Desk No.	Course & Section No.

Signature	Date	Witness/TA	Date

Note: Insert Divider Under Copy Sheet Before Writing

Exp. No.	Experiment/Subject	Date	
Name	Lab Partner	Locker/ Desk No.	Course & Section No.

Signature	Date	Witness/TA	Date

Note: Insert Divider Under Copy Sheet Before Writing

Exp. No.	Experiment/Subject		Date	
Name		Lab Partner	Locker/ Desk No.	Course & Section No.

Signature	Date	Witness/TA	Date

Note: Insert Divider Under Copy Sheet Before Writing

Exp. No.	Experiment/Subject	Date	
Name	Lab Partner	Locker/ Desk No.	Course & Section No.

Signature	Date	Witness/TA	Date

Exp. No.	Experiment/Subject	Date	
Name	Lab Partner	Locker/ Desk No.	Course & Section No.

Signature	Date	Witness/TA	Date

Exp. No.	Experiment/Subject	Date	
Name	Lab Partner	Locker/ Desk No.	Course & Section No.

Signature	Date	Witness/TA	Date

Note: Insert Divider Under Copy Sheet Before Writing

Exp. No.	Experiment/Subject		Date	
Name		Lab Partner	Locker/ Desk No.	Course & Section No.

Signature	Date	Witness/TA	Date

Note: Insert Divider Under Copy Sheet Before Writing

Exp. No.	Experiment/Subject		Date	
Name		Lab Partner	Locker/ Desk No.	Course & Section No.

Signature	Date	Witness/TA	Date

Note: Insert Divider Under Copy Sheet Before Writing

Exp. No.	Experiment/Subject	Date	
Name	Lab Partner	Locker/ Desk No.	Course & Section No.

Signature	Date	Witness/TA	Date

Note: Insert Divider Under Copy Sheet Before Writing

Exp. No.	Experiment/Subject	Date	
Name	Lab Partner	Locker/ Desk No.	Course & Section No.

COPY

Signature	Date	Witness/TA	Date

Note: Insert Divider Under Copy Sheet Before Writing

Exp. No.	Experiment/Subject		Date	
Name		Lab Partner	Locker/ Desk No.	Course & Section No.

Signature	Date	Witness/TA	Date

Exp. No.	Experiment/Subject	Date	
Name	Lab Partner	Locker/ Desk No.	Course & Section No.

Signature	Date	Witness/TA	Date

Note: Insert Divider Under Copy Sheet Before Writing

Exp. No.	Experiment/Subject		Date	
Name		Lab Partner	Locker/ Desk No.	Course & Section No.

Signature	Date	Witness/TA	Date

Note: Insert Divider Under Copy Sheet Before Writing

Exp. No.	Experiment/Subject	Date	
Name	Lab Partner	Locker/ Desk No.	Course & Section No.

Signature	Date	Witness/TA	Date

Note: Insert Divider Under Copy Sheet Before Writing